STUDENT UNIT GUIDE

NEW EDITION

CCEA AS Geography Unit 1
Physical Geography

Tim Manson and Alistair Hamill

PHILIP ALLAN FOR
HODDER
EDUCATION

Philip Allan, an imprint of Hodder Education, an Hachette UK company, Market Place, Deddington, Oxfordshire OX15 0SE

Orders
Bookpoint Ltd, 130 Milton Park, Abingdon, Oxfordshire OX14 4SB
tel: 01235 827827
fax: 01235 400401
e-mail: education@bookpoint.co.uk
Lines are open 9.00 a.m.–5.00 p.m., Monday to Saturday, with a 24-hour message answering service. You can also order through the Philip Allan website: www.philipallan.co.uk

© Tim Manson and Alistair Hamill 2014

ISBN 978-1-4718-0822-7

First printed 2014
Impression number 5 4 3 2 1
Year 2016 2015 2014

Cover photo: Tim Manson

Typeset by Integra Software Services Pvt Ltd, Pondicherry, India

Printed in Dubai

Hachette UK's policy is to use papers that are natural, renewable and recyclable products and made from wood grown in sustainable forests. The logging and manufacturing processes are expected to conform to the environmental regulations of the country of origin.

P2303

Contents

Content Guidance

Questions & Answers

Getting the most from this book

Questions & Answers

Fieldwork skills   Question **1A**

Question 1A Fieldwork skills

(a) Study the points below, which outline some important considerations made by a student when preparing for a geography fieldwork trip:
● Transport to the site
● Accessibility of the site
● Safety equipment
● Suitable clothing for fieldwork
● Communication devices

Select one of the planning considerations above and discuss its importance and role within your fieldwork.

(3 marks)

ⓔ 3 marks are awarded for an answer that deals with both the importance and the role of the selected factor and makes a convincing case, with appropriate reference to the individual fieldwork. 1–2 marks are given for a more simple response, which might fail to address either the importance or the role, with unconvincing reference to the individual fieldwork.

Student answer to question 1A

(a) When completing a fieldwork project in the sand dunes we had to think about specific safety equipment that would keep us safe on the site visit. We were worried about bad weather, so our teacher told us to make sure that we had a waterproof coat, waterproof trousers and a hat and gloves to keep us warm (it was a very cold, wet and windy day — so this advice was really important!). Also, we had to wear walking boots as the terrain was rough and this would give us more support. We brought waders and a safety throw rope as we were going to have to collect water from the sea to help with our experiment.

ⓔ 3/3 marks awarded   This makes reference to three different aspects of safety equipment that were used for this particular fieldwork. It discusses both the role of the equipment and the importance of this within the context of their visit to the sand dune.

(b) Describe and explain one sampling method that was used within your fieldwork.

(6 marks)

ⓔ You can select from a range of different sampling methods, but most relate to one of either random, systematic, pragmatic or stratified sampling. 3 marks are for a general description of the sampling method and the other 3 marks are for an explanation of how the sampling method works in the context of this particular fieldwork.

Unit 1: Physical Geography   63

CCEA AS Geography

About this book

Much of the knowledge and understanding needed for AS geography builds on what you have learned for GCSE geography, but with an added focus on geographical skills and techniques, and concepts. This guide offers advice for the effective revision of **Unit AS 1: Physical Geography** (including fieldwork skills), which all students need to complete.

The AS 1 external exam paper tests your knowledge and application of geographical skills and techniques and lasts 1½ hours. The unit makes up 50% of the AS award or 25% of the final A-level grade.

To be successful in this unit you have to understand:
- the key ideas of the content
- the nature of the assessment material — by reviewing and practising sample structured questions
- how to achieve a high level of performance within

This guide has two sections:

Content guidance — this summarises some of the key information that you need to know to be able to answer the examination questions with a high degree of accuracy and depth. In particular, the meaning of keys terms is made clear and some attention is paid to providing details of case study material to help to meet the spatial context requirement within the specification.

Questions and answers — this includes some sample questions similar in style to those you might expect in the exam. There are some sample student responses to these questions as well as detailed analysis, which will give further guidance in relation to what exam markers are looking for to award top marks.

The best way to use this book is to read through the relevant topic area first before practising the questions. Only refer to the answers and examiner comments after you have attempted the questions.

Content Guidance

Fieldwork skills

The first question on the AS Unit 1 paper asks you to make reference to some fieldwork/data collection that you have taken part in. Due to the diversity of fieldwork opportunities at A-level, it is impossible for this short guide to refer to every type of physical geography enquiry. Instead, it will cover one common example from the specification. However, the process and recommendations can be easily transferred to any piece of work.

When tackling any geographical enquiry, the first task is to identify the particular question and issues that are to be investigated.

The geographical fieldwork process

Most pieces of geography enquiry follow a similar investigation path:
- Title/aims/hypotheses
- Planning
- Data collection
- Data organisation (tabulation and presentation)
- Data analysis and data interpretation
- Drawing conclusions and evaluation

Examiner tip
Make sure you understand the sequence of this investigation path.

Planning for a fieldwork study

Many geography teachers are keen to involve students in the collection of primary data in order to develop their geographical skills.

Location selection

When planning a field trip, it is important to consider some key questions regarding the best location for your study:
- Is the location accessible?
- Is it easy to get to?
- Is it appropriate for a group of students to visit?
- Do we need permission to access the site?
- How much will it cost to get there or to get in?
- How much time will it take us to get there?
- Will it be safe?
- Could we damage the environment by being here?
- Is this the best example of this type of environment/feature?

Examiner tip
Apply these questions to your field trip and address any potential planning issues.

Risk assessment

No geography field trip can be undertaken in a hazard-free environment. Both students and teachers need to consider any risks before going into the field so that appropriate measures can be taken to ensure safety. For example, one risk in an area of sand dunes might be rabbit holes and undulating surfaces, which can be managed by advising students to wear sturdy shoes and to watch where they are walking.

Contingency plans need to be made to manage any risk. For school trips, a comprehensive risk assessment must be submitted to the Principal to demonstrate how any potential situation might be handled.

Examiner tip
Ask your teacher if you can see the risk assessment they had to submit before taking you on the field trip.

Pilot study

It might be important to conduct a pre-study site visit to make sure that the location is appropriate for the needs of the enquiry. You might need to check the accessibility, conduct a risk assessment and make sure that the results taken will allow the aims/hypotheses to be addressed. Equipment could be tested to make sure that it is appropriate for the study. If using a questionnaire survey, it might be good practice to test the questions a few times to make sure that they read well and make sense to the people you are testing.

Sampling

The purpose of fieldwork is to enable you to collect your own primary data, which might support any additional material gained from secondary sources.

- Primary sources/data comprise any new information that you have collected in the field. This might be done through observation or through measurement.
- Secondary sources/data comprise any new information that has been obtained from existing sources, such as maps, GIS, photographs or census data. Often in geography investigations you will need to use both primary and secondary information.

Knowledge check 1
What are some of the strengths and weaknesses of using primary data?

During field visits, sampling is needed because time restrictions make it impossible to study and take measurements from an entire area. Therefore, decisions have to be taken as to what is the most appropriate method of choosing which locations and aspects of the study are recorded.

The main sampling techniques are random, systematic, stratified (point, line and quadrat) and pragmatic.

Random sampling is when a random number table or random number generator app is used to give, for example, the sequence of people to ask, or houses to call in a street, when doing a survey.

Systematic sampling is when samples are taken using a pre-determined interval. For example, questionnaires might be completed for every 5th or 10th person who walks past, or soil studies might be taken every 5 m or 10 m along a survey line/transect.

Stratified sampling is a useful way of sampling when there are clear sub-groups within the dataset. For example, if conducting a sample of 70 questionnaires in a school population of 700 students, you might break the sample down so that those questionnaires allocated to a year group represent the size of that year group within the school. So, for example if there were 70 students in the sixth form they would get 10 questionnaires.

Examiner tip
Make sure that you know the differences between these sampling techniques. Questions commonly ask about the integrity of a sampling technique and how it was applied to your fieldwork.

Point sampling is when individual points are used within the investigation. For example, specific, accessible sites might be chosen within a river study.

Line or transect sampling is when a line is drawn on a map within an area and all data are collected along this line.

Quadrat sampling is when a piece of equipment called a quadrat is used to measure the amount of vegetation/type of vegetation or amount of ground coverage within an area. The quadrat is usually a square metal frame (the most common is 50 cm × 50 cm).

Pragmatic sampling is when decisions are taken to visit sites that are safe/accessible or which might demonstrate typical characteristics. Although this approach often allows for a simple fieldwork experience, it often introduces a huge amount of bias into the sampling technique.

Examiner tip
Consider the number of sites/questionnaires that you need for your sample. Often this is linked to the statistical technique that you select. For example, if you use Spearman's rank correlation you will need to visit at least 15 sites.

Fieldwork safety

Safety during fieldwork is very important and must be considered before going into the field (this will form part of any risk assessment strategy).

Weather conditions

Wet and freezing conditions can cause problems on field trips, while strong winds can be an issue when using fieldwork equipment. Wet clothes are uncomfortable and can cause rapid heat loss from a body, leading to hypothermia.

Strategy: Check the weather forecast and take appropriate clothing for the weather conditions and perhaps a change of clothes, a first aid kit and a thermal blanket.

Injuries

Many physical environments — for example, the coast, uplands, rivers and forests — can be dangerous places if care is not taken. Beaches, sand dunes and riverbeds have uneven surfaces. Slopes can be steep, uneven or have loose material that could cause a fall.

Strategy: Wear appropriate shoes/boots, avoid running, watch where you are going, carry a first aid kit with dressings and perhaps a bivi bag to carry an injured student.

Fieldwork equipment can be dangerous. For example, ranging poles, clinometers and metre sticks can be dangerous if carried in the wrong way. Some infiltration rings might have sharp edges, while safety ropes can cause 'rope burn'.

Strategy: Carry ranging poles with the spike down. All equipment should be checked to make sure there are no sharp edges or rust. Ropes should be rolled up carefully and carried responsibly.

Safety in urban areas

When working in urban areas, risks still exist in relation to traffic and other people. You should remain vigilant, ensuring that you do not get separated from your group.

Strategy: Use the green cross code when crossing roads, stay within groups, establish designated check-in times and places and have emergency contact numbers available.

Data collection techniques

The data collection techniques that you choose depend on the title and hypotheses that you have selected to investigate. Here (and in later sections) we will look at one example of physical fieldwork.

Fieldwork title: An investigation into seral succession of the psammosere at Magilligan Point

Hypotheses

1 Infiltration time increases with distance travelled from the sea.
2 Vegetation becomes more complex with distance travelled from the sea.
3 Ground cover increases with distance travelled from the sea.

Data collection methodology

In order to measure the changes to the sand dune system in relation to the distance from the sea, a systematic/transect sampling technique was used to collect results. Students started at the edge of the sea and had to move along a straight line, taking measurements at 15 sites at 15-metre intervals. This sample size was chosen as it is the minimum number needed to complete an accurate Spearman's rank correlation.

At each site, infiltration, vegetation characteristics and ground cover were measured.

Infiltration: An infiltrometer (infiltration ring — Figure 1), measuring cylinder, water and stopwatch were used to measure the rate of water infiltration at each site. The vegetation was cleared from the test area and 200 ml of water was poured into the infiltrometer, which had been pressed into the soil. The time taken for the water to clear into the soil was recorded using the stopwatch. This was repeated three times at each site to get an average time.

Figure 1 Using an infiltration ring on a sand dune

Examiner tip

Make sure that you have detailed knowledge of any equipment you use and can explain how it produced your results. Did you have any challenges in collecting your data?

Vegetation characteristics and ground cover: A 50 cm × 50 cm quadrat and species identification list were used to identify the type of vegetation at each site. The quadrat was 'tossed' randomly near the site and any vegetation type and the percentage of different types of ground coverage were noted on a results table.

Data organisation and presentation

The first step in sorting your data is to create a table of results. This should be brought into the exam as part of your prepared material.

Table 1 Results table for investigation into seral succession at Magilligan Point

Site (distance from the sea in metres)	Time (for 200 ml of water to infiltrate, in seconds)	Number of different species of vegetation	Vegetation cover %
1 (0)	10.1	0	0
2 (15)	32.3	0	0
3 (30)	62.5	1	20
4 (45)	95.5	2	35
5 (60)	120	4	50
6 (75)	133	2	55
7 (90)	62	1	65
8 (105)	156	5	80
9 (120)	132	6	90
10 (135)	171.4	6	95
11 (150)	235	7	95
12 (165)	245	5	95
13 (180)	266	7	90
14 (195)	256	8	95
15 (210)	262.5	8	100

Examiner tip

Make sure that you prepare your data table carefully and follow the instructions issued by the awarding body. The more organised your table is, the quicker you can answer the exam questions.

Examiner tip

The AS2 (Human Geography) Student Unit Guide gives further details on how to use statistical techniques in AS geography.

Examiner tip

You need to have practised one graphical representation and one statistical analysis. For this example I would suggest that you prepare Spearman's rank analysis and a scattergraph.

Some of the questions on the exam paper expect you to use the fieldwork table that you bring into the exam either to produce a **graph** or to apply a **statistical technique**.

Using just simple methods of statistical analysis (e.g. mean, median, mode and range) can cause more difficulties when trying to analyse and interpret the results later in the exam paper. Therefore, it might help to choose one hypothesis to which you can apply the Spearman's rank correlation.

For Hypothesis 1: Infiltration time increases with distance travelled from the sea — the following Spearman's rank table can be drawn from the results.

Table 2 Spearman's rank correlation table for fieldwork example

Site (distance from the sea in metres)	Rank	Time (for 200 ml of water to infiltrate, in seconds)	Rank	Difference in ranks, d	d^2
1 (0)	15	10.1	15	0	0
2 (15)	14	32.3	14	0	0
3 (30)	13	62.5	12	1	1
4 (45)	12	95.5	11	1	1
5 (60)	11	120	10	1	1
6 (75)	10	133	8	2	4
7 (90)	9	62	13	−4	16
8 (105)	8	156	7	1	1
9 (120)	7	132	9	−2	4
10 (135)	6	171.4	6	0	0
11 (150)	5	235	5	0	0
12 (165)	4	245	4	0	0
13 (180)	3	266	1	2	4
14 (195)	2	256	3	−1	1
15 (210)	1	262.5	2	−1	1

$\sum d^2 = 34$

The Spearman's rank formula is applied as follows:

$$r_s = 1 - \left(\frac{6 \sum d^2}{n^3 - n} \right)$$

$$r_s = 1 - \left(\frac{6 \times 34}{15^3 - 15} \right)$$

$$r_s = 1 - \left(\frac{204}{3360} \right)$$

$$r_s = 1 - 0.06$$

$$r_s = 0.94$$

An r_s result of 0.94 shows that there is a very strong positive correlation or relationship between the two variables (time it took for 200 ml to infiltrate through the soil and distance from the sea).

The graph and table in Figure 2 will usually be provided to help you to determine the significance of your result. In this case a critical value of 0.94 is within the 99.9% significant area. This means that the result is very significant and the relationship between the two variables can be commented on.

> **Examiner tip**
> Don't forget that this table uses information from the data table (Table 1) and cannot be pre-prepared.

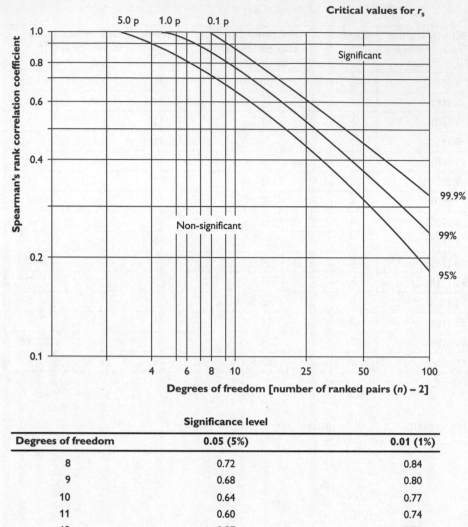

Figure 2 Spearman's rank correlation significance graph and table

The graph shows:

Critical values for r_s

Y-axis: Spearman's rank correlation coefficient (0.1 to 1.0)

X-axis: Degrees of freedom [number of ranked pairs (n) – 2] (4, 6, 8, 10, 25, 50, 100)

Curves labelled: 5.0 p, 1.0 p, 0.1 p, with significance levels 99.9%, 99%, 95%

Regions: Significant, Non-significant

	Significance level	
Degrees of freedom	**0.05 (5%)**	**0.01 (1%)**
8	0.72	0.84
9	0.68	0.80
10	0.64	0.77
11	0.60	0.74
12	0.57	0.71
13	0.54	0.69
14	0.52	0.67
15	0.50	0.65
20	0.47	0.59

Other questions in relation to data presentation might ask you to use data from the table to draw a graph that is relevant to the aim of the fieldwork. Make sure that you draw a graph that allows the opportunity for a full analysis in later questions. For Hypothesis 1: infiltration increases with distance travelled from the sea, a scattergraph (Figure 3) or a line graph is appropriate.

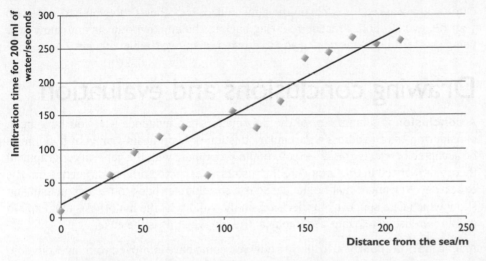

Figure 3 Scattergraph to show time taken for water to infiltrate through a sand dune against the distance from the sea

Care must be taken when drawing the graph as marks are awarded for the accuracy of your presentation. In previous exam series, 7 marks were awarded for a graph, with 1 mark for a specific and accurate title, 2 marks for the use of mathematical conventions (labelled axes, use of a key or scaling of the graph), 3 marks for accuracy (the precise plotting of values) and 1 mark for the method in selecting an appropriate graphical technique in relation to the aim and data table.

Data analysis and data interpretation

After drawing your graph or using a statistical technique on your data, you then need to analyse or interpret your findings with reference to your stated aim.

Data analysis

You might be asked to *describe* the data, so you should practise writing about the *patterns and trends* that you notice in the results. Use figures from the graph/table/ statistical technique to add weight to your description. Note the highs/lows and any averages that you have calculated. Describe any relationships on the graph — such as positive or negative correlations — and comment on the significance of the result. Describe any unusual results.

Data interpretation

You might be asked to *explain* or *interpret* the data, so practise making reference to the results shown in the graph/statistical technique and then try to explain the reasons behind your results. What are the factors that have controlled or created this situation? Do your results help you to prove or disprove the hypothesis that you have been testing? How does this fit within general geographical theory? Is this what you expected? Are there results that go against what you expect?

Practise analysing and interpreting your graph, and statistical analysis. These are common questions on the exam paper.

Often you are asked to reflect on the aims and/or hypothesis that you had stated on your fieldwork report. Practise referring back to the aim/hypothesis and show how using this technique has helped you to prove or disprove the hypothesis.

Drawing conclusions and evaluation

A **conclusion** is a summary of the information and evidence that you have been considering as you address a particular question or hypothesis. Can your hypothesis be accepted or rejected? How do your findings compare with the general geographical theory associated in this area? Are the results as expected or very different from the expected? Why might that be the case? You should refer back to the hypothesis and show which data support your decision on the validity of the hypothesis and explain (with reference to relevant geographical theory) why this is the case.

An **evaluation** allows you to discuss how you could have improved your investigation. What are the main strengths and weaknesses of the investigation? What modifications could you have made to improve the accuracy of your fieldwork? How could your investigation have been further extended?

How could you have improved your data collection methods?

Was your methodology accurate enough to produce results that you could trust? How could you have gone further to make sure that your results were accurate? What additional equipment might you have used to get more accurate results?

How could you have improved your sampling technique?

Was it fit for purpose? Would a different sampling technique have allowed more accurate results? Did you visit too many/too few sites?

How could you have improved your conclusions?

Are your conclusions reliable and accurate? Were the title/aim/hypotheses that you selected appropriate for what you wanted to study? Was the location of your study appropriate?

Preparing the fieldwork report and table

Remember that you are expected to submit a fieldwork report and data table to your teacher or examinations officer before the examination.

The **fieldwork report** (a short report of around 100 words) should include:
- a general title for the fieldwork
- a brief statement of the aims/purpose/issues that provide the theoretical context for the personal investigation element of the fieldwork (this can be key questions/hypotheses)
- a brief outline of the spatial context of the study (including a location map, if desired)

The report should not address any other aspects of the investigation.

The **data table** (an attached table of data) should include:

- a specific title
- data collected for all variables relevant to the proposed aim/purpose of the study in the report
- primary/secondary data essential to the aim
- quantitative data (numbers), essential to allow graphical representation and statistical analysis
- adherence to normal conventions (all variables stated clearly and precise units of measurement)
- raw data (averages or other statistical calculations should not be included)

Examiner tip

Get your finished written report and data table to your geography teacher a few days before the examination and make sure that you fill in and sign the yellow declaration sheet.

Summary

- In preparation for the fieldwork skills question on the paper (Question 1) you need to bring a fieldwork report and a data table into the exam with you.
- Make sure that you understand the usual order of investigation within the fieldwork process.
- Geography field trips require careful planning, risk assessment and measures to ensure health and safety at all times.
- A variety of data collection techniques can be used to observe and measure geographical data.
- Most fieldwork will involve collection of primary data.
- You should consider the best sampling technique to ensure the integrity of your recorded data.
- A statistical technique such as Spearman's rank analysis should be practised and applied to an aspect of the data table.
- A graphical technique such as a scattergraph should be practised and applied to an aspect of the data table.
- Practise referring to your hypothesis/aim to prepare for questions on data analysis, data interpretation, conclusions and evaluation.

Topic 1 Rivers

Processes and features in fluvial environments

The drainage basin as a system

A drainage basin is the area of land that is drained by a river and its tributaries. As water falls on the land as precipitation, gravity pulls it downhill and back towards the sea. The boundary of a drainage basin is known as the watershed. This is in the form of a ridge of high land so that any water that falls inside the watershed line will drain through the system. Water that falls outside this boundary will flow through a different drainage system.

Components of a geographical system

The systems approach gives us a very useful framework for analysing and understanding the drainage basin (Figure 4). An open system will have four main elements: **inputs** into and **outputs** from the system (of both energy and matter) and **stores** and **transfers** within the system.

Examiner tip

Make sure that you understand the differences between the components of an open system and apply them to the different processes of the drainage basin system.

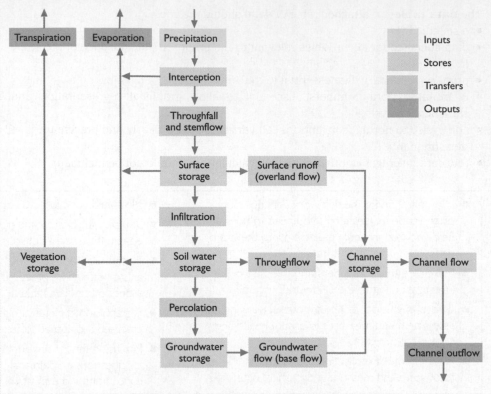

Figure 4 The drainage basin as a system

Examiner tip

When learning these terms, picture the connections between them. Imagine the journey of a raindrop as it transfers between each of the stores.

Components of the drainage basin

1 **Precipitation (input):** Water falls from clouds (as drizzle, rain, sleet and snow) towards the surface.

2 **Interception (store):** The precipitation is caught and held for a short time by vegetation before it reaches the soil store. When there is more vegetation, more interception will occur and there is the potential for more evaporation, especially in the summer.

3 **Stemflow and throughfall (transfers):** The movement of water from the interception store to the surface store, either by flowing down the stems/branches (stemflow) or dripping off the leaves (throughfall).

4 **Surface store:** Water is stored temporarily on the surface (e.g. in puddles).

5 (a) **Infiltration (transfer):** Water *enters* the soil from the surface store.

   (b) **Overland flow/surface runoff (transfers):** Water runs over the surface of the land following a rainstorm. This occurs when the soil either becomes *saturated* (because too much rain has fallen and there is no more room in the soil store) or because the rain intensity is too high and *exceeds the infiltration capacity* (i.e. the maximum rate at which water can infiltrate into the soil). It also occurs on *impermeable surfaces*, for example tarmac in urban areas.

6 **Soil store:** Water that has infiltrated is stored in the surface layers of the soil before experiencing either throughflow or percolation.

7 **Throughflow (transfer):** Water moves downhill through the soil, close to and parallel with the surface, to the river.

8 **Percolation (transfer):** Water moves *further down into the soil* from the soil store to the groundwater store.

9 **Groundwater store:** The permanent store of water in the lower layers of the soil and the bedrock.

10 **Groundwater flow (transfer):** The movement of water from the groundwater store in the lower layers of soil and the bedrock to the river.

11 **Evaporation (output):** Water is changed into water vapour from various stores such as interception and surface storage. The main factor affecting the rate of evaporation is temperature.

12 **Transpiration (output):** Water vapour is taken from vegetation and plants into the atmosphere. This is affected by factors such as the vegetation type (e.g. deciduous trees lose their leaves in winter to reduce transpiration) and moisture availability.

What factors affect transfers and stores of matter in a drainage basin?

Vegetation: Thick vegetation (like forests) can affect the flow of water as it creates more opportunities for interception and for evaporation from the leaves of the trees. Any water that does not directly reach the ground will take a much longer time to go through the system. Soil within dense forests is unlikely to reach infiltration capacity.

Soil type: Some soils are *porous* (they have big spaces, e.g. *sandy soils*) and this allows water to infiltrate quickly. Other soils are less porous (e.g. *clay soils*) and water cannot infiltrate, making overland flow the main transfer.

Seasons: The warmer temperatures of summer encourage more evapotranspiration, thus lowering discharge levels in rivers. If the soil becomes hard-baked it will not allow infiltration. When deciduous leaves fall in the autumn there is less interception. In the winter the ground could be frozen, preventing infiltration.

Geology: If the rock underlying the soil is porous, then there is more percolation and groundwater flow. This reduces the soil store and the amount of water transferred by overland flow.

Urban areas: These areas have *less vegetation* so there is less interception and less infiltration due to the amount of *impermeable surfaces*. Water is often channelled directly and quickly into rivers via *drains and sewers*.

Relief: In steeper drainage basins, less infiltration tends to occur and so *surface runoff* dominates and there is less water in the soil store. Conversely, in lowland areas more infiltration can occur and water tends to remain in the soil store for longer before being transferred via throughflow.

How heavy the rain is: During heavy, intense rainstorms, rain can fall at a rate that is faster than the soil can allow it to infiltrate (the maximum rate of infiltration is called the **infiltration capacity**). If this occurs, overland flow is likely to happen.

Examiner tip

Top-level answers must use a good range of terms. Be familiar with these terms and practise using them in your answers.

Knowledge check 2

Distinguish between infiltration and percolation.

Examiner tip

If an exam question asks for impacts of these factors on both *stores* and *transfers*, make sure both are clearly and explicitly covered in your answer. Questions can also explore what happens to stores/transfers if *land use changes*, for example if a rural area becomes urbanised.

Examiner tip

Sediment is also transferred through drainage basins. See p. 21 for more information on this.

Knowledge check 3

How do stores and transfers vary throughout the year in a deciduous forest in the UK?

Discharge and the storm hydrograph

Discharge is the amount of water that passes a particular point in the river per second. It is measured in cubic metres of water per second (cumecs) and is calculated by multiplying velocity by cross-sectional area.

Discharge in a river is not constant and rises and falls over time, especially in response to a rainstorm. These changes in flow can be plotted on a time/discharge graph called a **storm hydrograph** (Figure 5).

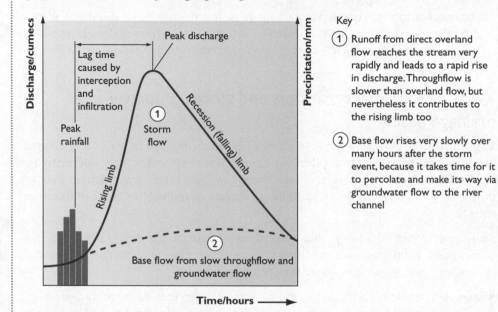

Figure 5 Storm hydrograph

The hydrograph has two main elements:
- **Base flow** is the 'background' flow contributed slowly and steadily by groundwater flow.
- **Storm flow** is the additional flow contributed by precipitation.

Note also the following terms: **peak discharge** (the maximum discharge level), **lag time** (the time difference between peak precipitation and peak discharge), **rising limb** and **falling (recessional) limb**.

As the rainstorm begins, the discharge rises very slowly. There is little initial change during the rainstorm as most rain does not fall directly into the river. So it experiences interception, infiltration, throughflow and so on, all of which take time.

Shortly afterwards, the **rising limb** of the discharge rises steeply towards its **peak discharge**. This is because the storm water has now started to reach the channel after experiencing infiltration and throughflow. If the rain is intense and/or if the soil becomes saturated, water arrives via overland flow too.

After reaching its peak, the **falling limb** falls more slowly as the main flow now is throughflow, which is slower than overland flow.

Finally, the discharge returns to its original level. The influence of the storm flow has now passed and the discharge goes back to **base flow**, which is the steady feed of water into the river via groundwater flow.

Factors affecting discharge and the storm hydrograph

The hydrograph in Figure 5 shows a model response (i.e. a typical or average response) to one single rainstorm. In reality, there are many factors that can *modify* this model response and *affect the shape* of the hydrograph. We will consider two contrasting types (Figure 6): flashy hydrographs (which have a short lag time and a high peak discharge) and flat hydrographs (with longer lag times and lower peak discharges).

Examiner tip

Exam questions often require you to show understanding of how the model storm hydrograph can be affected by these various factors. Make sure you can explain *how* and *why* each factor can influence the hydrograph. How are the peak discharge and lag time affected?

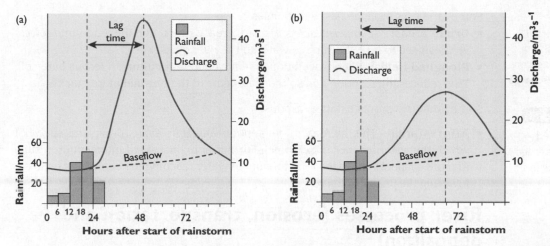

Figure 6 Flashy (a) and flat (b) hydrographs

The nature of the storm

- **Intensity and length of precipitation:** More intense rain causes the soil to become saturated and/or the infiltration capacity to be exceeded, so more overland flow occurs. A similar situation can occur if the rain is prolonged and the soil store becomes filled. This produces the shorter lag time and higher peak discharge of the flashy hydrograph.

Nature of the drainage basin

- **Basin size:** In smaller basins, the precipitation has less distance to travel before it reaches the mouth, so the hydrograph will be shorter and steeper.
- **Basin shape:** Shorter, more rounded basins are more likely to be flashy as the water from the basin tends to arrive more quickly at the mouth. In a longer, thinner basin, the water that falls near the source has much further to travel and so it will produce a flatter hydrograph.
- **Basin relief:** In steeper basins, under the influence of gravity, water will make its way to the mouth more quickly.

Examiner tip

As we look at these factors, we will discuss how they produce flashy hydrographs. To work out how they produce flat hydrographs, simply state the opposite. For example, if in steep basins water arrives at the mouth more quickly, producing the shorter lag time and higher peak discharge, then in gentle basins water arrives more slowly, producing a longer lag and lower peak.

Source Where drops of water join to start a river.

Mouth Where the river flows into the sea.

- **Soil type:** Clay soils have much smaller pore spaces and so do not allow for much infiltration. As a result, overland flow is more likely and so the water reaches the channel quickly. The opposite is the case for sandy soils.
- **Geology:** Some rocks, such as basalt, are impermeable and less infiltration occurs. This produces flashier hydrographs. More permeable rocks, such as chalk, allow infiltration and produce flatter hydrographs.
- **Drainage density (number of streams per km²):** Drainage densities are higher in areas with clay soils. As the water does not have to travel far to get to a channel, it travels to the mouth more quickly. This tends to produce shorter lag times and higher peak discharges.

Land use

Some land uses produce flashy responses:

- **Urban areas:** The *impermeable surfaces* increase runoff and the *drains and sewers* are designed to take the surface water to the river quickly.
- **Ploughed fields:** Where vegetation is removed for agriculture, it leaves bare soil. This reduces interception and so the water gets to the channel more quickly.

Some land uses produce flatter responses:

- **Afforestation:** This increases interception and thus slows down the speed at which the water reaches the channel. Furthermore, increased interception results in more evaporation, so the total amount of water reaching the channel is reduced, lowering the peak discharge.

River processes (erosion, transportation and deposition)

Erosional processes

Erosion is the wearing away of the bed and banks in the river. There are four main erosional processes:

- **Abrasion/corrasion** is the most effective form of river erosion and occurs when the river uses its load to erode the bed and banks by scraping and scouring. It is particularly effective at times of higher discharge when the river has enough energy to transport larger particles (see the Hjulström curve on p. 21).
- **Hydraulic action** occurs due to the physical force of the water against the bed and banks. On the outside bends of meanders, for instance, the currents push water into cracks, causing pressure that leads to erosion. Hydraulic action is more effective in rapids and at waterfalls. It tends to move unconsolidated sands and gravels on the riverbed.
- **Corrosion/solution** is the dissolving of soluble materials in the bed and banks by weak acids in river water. This is a chemical reaction rather than a physical process and so is not dependent on the energy levels in the river. It is most effective in rocks containing carbonates, such as limestone.
- **Attrition** occurs when the load particles come into contact with other load particles and the bed and banks. As a result the rough edges are smoothed and the particles becomes smaller and more rounded (this is particularly noticeable as you move downstream).

Examiner tip

Remember to connect clearly and explicitly each of these factors with how they affect (1) the **lag time** and (2) the **peak discharge**.

Knowledge check 4

How might increased urbanisation in a drainage basin affect lag time and peak discharge?

Knowledge check 5

Which of the erosional processes are affected by changes in river energy levels as discharge increases or decreases?

Erosional processes can occur in two directions:

- **vertical**, creating river valleys — this is more common in the upper course of the river
- **lateral** (horizontal) as meandering rivers widen floodplains

Transportation processes

Any energy not lost by the river by friction can be used to transport sediment. There are four types of transportation.

- **Suspension:** The smaller particles of clay, silt and sand can be carried along by the turbulence of the river. This tends to be the most effective form of transportation and it explains why rivers in their lower course tend to be brown in colour.
- **Solution:** The material eroded by corrosion is carried along and dissolved in the water. This form of transportation can be significant in limestone areas, but tends to be less important in other areas.
- **Saltation:** The smaller bedload, such as pebbles and gravel, can be bounced along the riverbed by turbulence during times of higher discharge.
- **Traction:** The largest boulders in the river can be rolled along the riverbed during times of very high discharge.

Deposition processes

When river energy drops, deposition occurs in various places in the river such as the inside bends of **meanders** (these are called **point bar deposits**), on **floodplains** as a river overflows its banks and in **deltas**.

Examiner tip

In the exam, if you need to expand on your explanation of deposition, refer to the details of how deposition occurs in meanders, floodplains and deltas (pp. 23–26).

The Hjulström curve

The Hjulström curve (Figure 7) shows the velocity needed to erode, transport or deposit different sized particles.

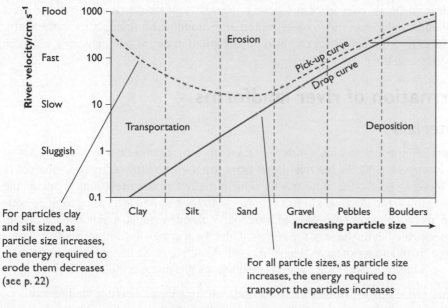

Figure 7 The Hjulström curve

The Hjulström curve shows two lines:

- the **critical erosion velocity/pick-up curve** (i.e. the speed needed to pick up the particles)
- the **critical deposition velocity/drop curve** (i.e. the speed at which there is no longer enough energy to transport the particles)

For most of the graph, the more energy a river has, the greater its ability to erode and transport particles. So, as the river's energy increases, it is able to erode and transport increasingly large particles. This is because *larger particles with greater mass require more energy both to erode and transport them*. For example, gravel will be eroded at velocities around $13\,cm\,s^{-1}$, whereas boulders require velocities closer to $500\,cm\,s^{-1}$ before they will be eroded.

There is one section of the graph where this relationship does not hold, however. For particles of silt or clay, the smaller the particle size, the greater the energy required to erode them. For example, velocities need to be up around $300\,cm\,s^{-1}$ again before the smallest clay particles will be eroded. There are two main reasons for this.

- Clay particles are **cohesive** (they stick together) and so need more energy to separate them one from another. Once separated, however, they are so small that they are easily transported.
- Clay particles form smooth surfaces when packed together. This means that, as the river water flows over them, there is less turbulent flow, and so fewer eddies in the water that could scour off particles from the surface. In contrast, particles that are sand sized and larger encourage these eddies.

In all cases, the velocities required to erode are greater than those required to transport. However, for particles gravel sized and bigger the difference between the erosional and depositional velocities is quite small. This means that larger particles are deposited soon after they have been eroded if river velocities drop. For the smallest particles, river velocities have to be very low for deposition to occur. As a result, silts and clays tend to be deposited after flooding on floodplains as the water infiltrates into the soil. They can also be deposited in deltas and estuaries, aided by the process of flocculation (see p. 26).

Formation of river landforms

Waterfalls

Waterfalls form where bands of harder rock (such as basalt) overlay bands of softer rock (such as chalk). As the river flows from the harder to the softer rock, the softer rock tends to be eroded more quickly. This initially forms a **step**, which causes the water to fall vertically, thus losing its contact with the bed and increasing its speed due to the drop in friction. Over time, erosion (particularly hydraulic action and corrasion) is concentrated at the base of the waterfall and it excavates the softer rock to create a **plunge pool** and a **notch** as the harder rock is undercut. As the notch grows in size, the overhanging harder rock above it collapses, as it is unsupported.

By this process, the waterfall tends to migrate upstream, leaving behind a steep-sided **gorge** (Figure 8). The **Gulfoss waterfall in Iceland** has a particularly good example of a gorge.

Examiner tip

Take time to understand *why* the relationships shown in the Hjulström curve exist, especially the key principles relating to particle *size* and particle *cohesion*.

Examiner tip

Questions on river features sometimes require annotated diagrams. You can earn full marks for a clear, fully annotated diagram that incorporates explanations. Practise drawing *and* annotating all the diagrams in this section. The success criteria for diagrams are accuracy, clarity and detailed annotation.

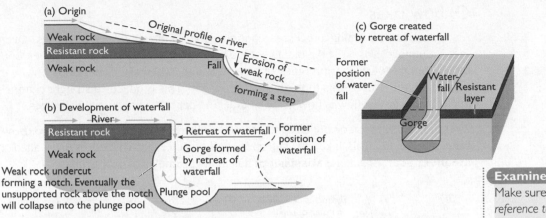

Figure 8 Waterfall formation

Pools and riffles/meanders

Along rivers carrying particles that are generally sand size and larger you will often have alternating features called **pools** (deeper areas with smaller particles such as silt) and **riffles** (shallower areas made up of larger particles such as gravel or pebbles). Due to the increased frictional drag caused by the larger particles in the riffles, the maximum velocity flow tends to swing to avoid these areas. This creates side-to-side motion within the water and is the start of the process of meander development.

Once this side-to-side motion is established, it produces variations across the channel. As the fastest flow swings to avoid the riffles, erosion (especially corrasion and hydraulic action) is concentrated on one of the banks. This moves it back laterally, deepening the channel here and causing the undercutting of the river bank, producing a steep-sided **river cliff**. At the same time, on the other bank where river velocities are lower, deposition occurs, forming a gently sloping bank made up of **point bar deposits**. The slope down towards these deposits is known as a **slip-off slope**. Over time, as erosion continues on the outside of the bend and deposition on the inside, the river becomes more **sinuous** as it moves laterally across the floodplain (Figure 9).

As well as moving laterally, meanders tend to migrate downstream. This is because the fastest flow does not follow the precise shape of the channel and thus the point of maximum erosion is just downstream of the halfway point of the meander bend. The **Mississippi River in the USA** has many large meanders along many kilometres of its course.

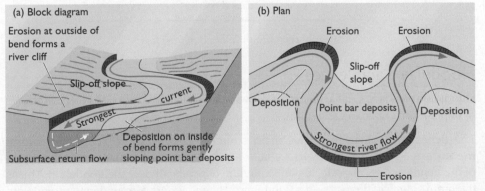

Figure 9 Meander in (a) cross-section and (b) plan view

Knowledge check 6

Describe the cross-sectional shape of a meander.

Oxbow lakes

As the meander forming processes continue, the river's sinuosity increases. Over time, continued erosion of the outside of the bend causes the gap in the meander loop to become narrower. During times of high discharge when the river floods onto the floodplain, the river breaks through the narrow gap. This produces a straight channel and thus the fastest flow is now in the middle and deposition occurs at the edges.

Over time, this deposition builds up to cut off the old meander loop, creating an **oxbow lake** (Figure 10). This lake can silt up, forming a crescent-shaped marsh called a **meander scar**. Again, the **Mississippi** has many examples of oxbow lakes.

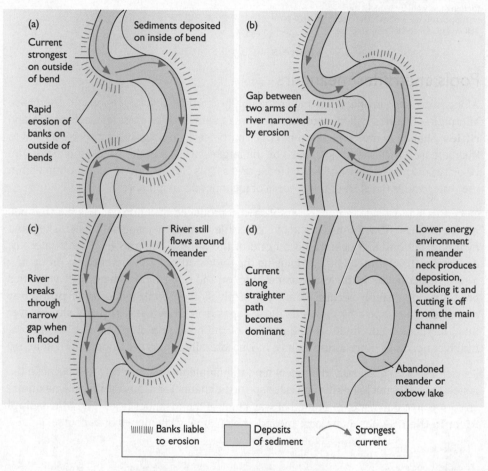

Figure 10 Formation of an oxbow lake

Floodplains and levées

Floodplains are large, flat areas around lowland rivers characterised by large amounts of deposition (known as alluvium).

The deposition comes from two main sources, the first of which is **migrating meanders**. As the meanders weave laterally, they deposit **point bars** on the inside of the bends as they migrate, leaving alluvium deposits all over the floodplain. This is the main source of deposits on a floodplain. The meanders also widen the floodplain by lateral erosion at the outside of their bends, often leaving prominent slopes called **bluff lines** at the edge of the floodplain (Figure 11).

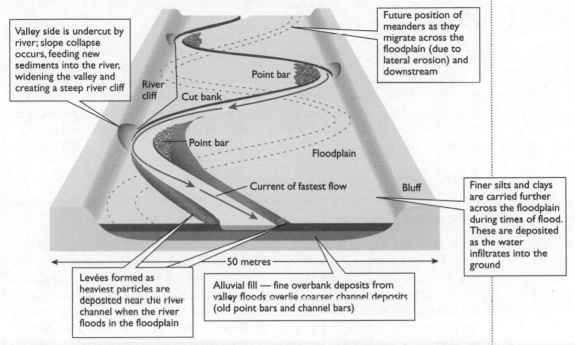

Valley side is undercut by river; slope collapse occurs, feeding new sediments into the river, widening the valley and creating a steep river cliff

Future position of meanders as they migrate across the floodplain (due to lateral erosion) and downstream

Point bar

River cliff

Cut bank

Point bar

Floodplain

Current of fastest flow

Bluff

Finer silts and clays are carried further across the floodplain during times of flood. These are deposited as the water infiltrates into the ground

50 metres

Levées formed as heaviest particles are deposited near the river channel when the river floods in the floodplain

Alluvial fill — fine overbank deposits from valley floods overlie coarser channel deposits (old point bars and channel bars)

Figure 11 Formation and characteristics of a floodplain

Secondly, deposition occurs as a result of **river flooding**. If the river overflows its banks, it experiences a significant increase in friction (due to contact with the ground and with floodplain vegetation). This lowers the velocity, encouraging deposition.

The Hjulström curve tells us that, as energy drops, the largest particles (sands and gravels) are deposited first. They tend to form low ridges along the edges of the river called **levées**. Meanwhile, the finer silts and clays are carried further away from the river across the floodplain and are deposited as the river water slowly infiltrates into the ground over time.

As the levées increase in height over time, and if there is further deposition on the riverbed, this can result in the river flowing at a higher level than the floodplain. In some places, such as the **Mississippi**, river engineers have increased the height of the natural levées to increase channel capacity and attempt to reduce the flood risk.

Deltas

Deltas are depositional features extending from the mouth of the river into the sea or a lake. The river channel splits and divides into distributaries, which weave through the delta.

Distributary A branch of a river that flows out from the main channel.

Factors affecting delta formation

Deltas form under the following conditions:
- **When the rate of deposition exceeds the rate of erosion:** Deposition rates are higher when a river has a large amount of sediment, for example a large river such as the **Nile**. Erosion rates are lower when the marine environment has a smaller tidal range and weaker currents.

- **Flocculation:** When fresh water mixes with seawater, chemical reactions with the salt cause clay particles to coagulate (stick together). This increases their weight and increases depositional rates.
- **A gentle sea floor gradient:** This increases deposition and aids delta formation.

Delta characteristics

As the sediment leaves the river mouth, it tends to be deposited in three layers: topset beds, foreset beds and bottomset beds (Figure 12a). As the river carries its sediment load out into the open water and the energy levels drop, the Hjulström curve tells us that the heavier sands and silts are deposited first while the lighter silts and clays are carried further out.

Delta types

Depending on the overall energy environment, the plan view pattern that can be formed is one of two types.

Arcuate deltas (Figure 12b) grow out from the coastline with a convex outer edge. One of the best examples of this is the delta of the **River Nile**. The deposition at the mouth of a river is usually triangular in shape. This is due to the blocking of the river mouth, which forces the river to break into a number of other channels called distributaries at angles to the original course. The distributaries weave back and forth, depositing sediment, which, over time, forms the characteristic fan shape of the arcuate delta. In addition, wave erosion tends to smooth off the outer edge, giving the arcuate delta a more pronounced fan-like shape.

Bird's foot deltas (Figure 12c) occur where delta formation is more river-dominated and less subject to tidal or wave action. The deposition pattern appears more haphazard and the delta can take on a multi-lobed shape that resembles a bird's foot. An example of a river-dominated delta is the **Mississippi River** delta.

Knowledge check 7

Describe **one** factor that can determine the rate at which a delta forms.

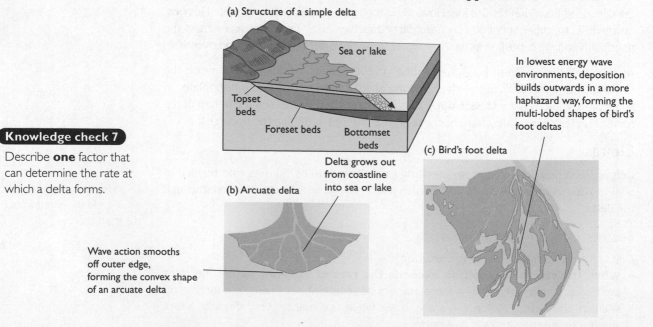

(a) Structure of a simple delta

Sea or lake

Topset beds

Foreset beds

Bottomset beds

Delta grows out from coastline into sea or lake

(b) Arcuate delta

Wave action smooths off outer edge, forming the convex shape of an arcuate delta

In lowest energy wave environments, deposition builds outwards in a more haphazard way, forming the multi-lobed shapes of bird's foot deltas

(c) Bird's foot delta

Figure 12 Deltas

- A drainage basin is an open system consisting of inputs (especially precipitation), stores (such as interception and soil store), transfers (such as overland flow and infiltration) and outputs (including evaporation and transpiration).

- The ways in which the stores and transfers operate can be influenced by a number of factors including vegetation, soil type and seasons. Land use change (such as the change from vegetation to urban land uses) can also affect the stores and transfers.

- The changes in a river's discharge following a storm can be shown in a storm hydrograph, which shows peak discharge and lag time.

- Storm hydrographs can be flashy or flat depending on a range of factors such as the nature of the

rainstorm (amount and intensity), characteristics of the drainage basin (basin size, relief, drainage density) or land use (vegetated or urbanised).

- Rivers erode, transport and deposit sediment by a variety of processes (corrasion, hydraulic action, corrosion and attrition; suspension, solution, saltation and traction).

- The different velocities needed to erode, transport or deposit particles of differing sizes are shown on the Hjulström curve. As particle size increases, the energy needed for erosion and transportation increases. However, for the smallest particles of silt and clay, due to their cohesive nature, more energy is required to erode them, even though they are smaller.

Human interaction with the fluvial environment

Case study

Causes and impacts of flooding in the Mississippi basin, 2011

Physical and human causes of the 2011 flood

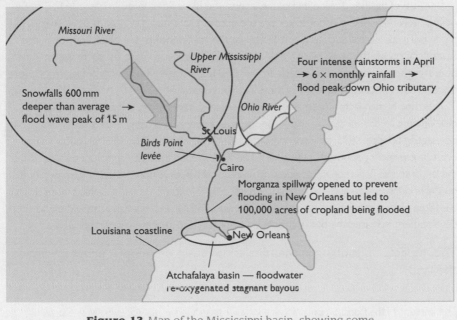

Figure 13 Map of the Mississippi basin, showing some of the causes and impacts of the 2011 flood

Examiner tip
Refer to this map as you read through the case study to give yourself a good sense of where all these places are in relation to each other.

Tributary A small river or stream that feeds into the main channel.

Confluence Where two rivers meet and join.

Physical causes were the primary trigger of the devastating floods of 2011 in the Mississippi basin in the form of two sets of unusually extreme weather events happening at similar times in different parts of the drainage basin.

Firstly in the upper Mississippi and Missouri tributaries to the west (Figure 13), the *snow falls* of winter 2010/11 were record breaking, with the snow being around 600 mm deeper than average. This meant that during the spring snow melt in April there was a higher quantity of water entering the river system. This raised the peak discharges and sent a flood peak down these tributaries that peaked at 15 m in Thebes, Illinois (just to the south of the confluence of these two rivers).

At the same time, the Ohio tributary to the east was also having record-breaking weather, but in this case in the form of a series of *four intense rainstorms* in April, which produced six times the average monthly rainfall for the area. This also sent a flood peak down this tributary towards the confluence of the Ohio with the Mississippi at Cairo, Illinois.

Once these physical factors had triggered the increased flood risk, various **human causes** contributed to increasing the scale of the floods. The first of these was related to the human management strategies used along the Mississippi, especially the **levées**. Following the devastation of the 1927 floods, this hard engineering strategy was used extensively throughout the river basin along around 3,000 km of the river. Although levées have been successful to an extent in reducing flooding (for example, the 450 km of levées in the St Louis district are estimated to have prevented over $11 billion in damage), they can also increase flooding. By stopping the water from spreading laterally over the floodplain, they tend to raise peak discharges and increase river velocities. For instance, the 2011 flood peak in St Louis was 3 m higher than the same volume would have been in 1927 prior to the extensive levée building. These higher and faster discharges put more pressure on the levées and can cause them to fail, funnelling the floodwater into the unfortunate areas around the failure point.

Because of this flood risk effect of the levées, the Birds Point levée in Missouri had to be deliberately breached in order to reduce the erosional pressures on levées downstream in more high-value areas such as Cairo, Illinois (at the confluence of the Mississippi and Ohio Rivers). Although the levée breach did lower the river levels in Cairo by 30 cm within a few hours, in the area around the breach around 100 homes and 13,000 acres of farmland were flooded as a direct consequence of these human actions.

A second human factor that makes flooding worse is urbanisation, and especially *urban development on floodplains*, which puts people and property at greater risk. Following the devastating floods in 1993, the Federal Emergency Management Agency (FEMA) adopted more of a soft engineering approach, buying back 12,000 homes on the Mississippi floodplain to convert the land into safe flood zones. However, pressure for urban development led to increased building on the floodplain. Amazingly, according to Professor Kasky from St Louis University, more building has taken place in the floodplain around St Louis since 1993 than before this date. All of this new development increased the demand for levées, which in turn increased overall flood risks.

The impacts of flooding in the Mississippi basin on people, property and the land

People

The impacts of the flood on people were *largely negative*. The USACE (US Army Corps of Engineers) estimated that more than 43,000 people felt some effects of the flooding. Despite the huge scale of the flood, the number of fatalities was quite low overall, largely due to the excellent forecasting and evacuation plans. *Twenty people lost their lives in total*, including eight in Arkansas.

Examiner tip

If you are asked to evaluate whether physical or human causes are more responsible for causing flooding, you should say that *extreme physical events* are the main *trigger* of flooding but that *human factors* can make the *scale* of the flooding worse.

People were also *displaced from their homes due to the flooding*. For example, when the USACE blew a 3 km hole in the Birds Point levée to protect the city of Cairo from flooding, around 200 residents in New Madrid County were forced to leave their homes. In addition, the USACE opened the Morganza spillway north of New Orleans for the first time in 38 years to lower the flood peak heading for the city; however, the water that flooded the surrounding land put 25,000 people at increased flood risk, including the 12,000 residents of Morgan City, who had to try to protect their homes with sandbags.

People were also affected by the *economic impacts of the floods*. Farmers in particular were badly hit with agricultural loses due to flooded land, including crops estimated at $800 million in Mississippi, $500 million in Arkansas and $320 million in Memphis. In addition, the 19 casinos along the Mississippi were closed down for around 6 weeks, resulting in loss of income for the 13,000 employees and $14 million in lost tax revenue). The *only positive impact* of the flood was an increase in fertiliser and seed sales for companies such as Monsanto and Dupont, as farmers needed to replace the crops lost by flood damage.

Property

The impacts of the floods on *property* were also *very negative*. The USACE estimated that the total damage to buildings was around $2 billion and that more than 21,000 buildings were affected. Some of this damage resulted from human attempts to prevent damage in other, higher-value areas. For example, when the Birds Point levée was blown up to save Cairo, the floodwaters spilled out onto the surrounding landscape and 100 homes were flooded as a result. Further south, in Memphis, the levées that had been built to protect the downtown area worked, but the floodwaters instead covered the suburban areas, flooding 1,300 homes.

Land

The impact of the flooding on the land was *partly negative and partly positive*. For example, when the Morganza spillway was opened in May to alleviate the flood risk in New Orleans and Baton Rouge, around 100,000 acres of cropland, some planted with sugar cane and rice, were flooded. Around 2.2 million acres of land were affected in the delta region, which is around 1% of the entire US cropland area.

However, there were some *positive impacts of flooding* on the land. One example was in Louisiana, where the oil companies had built a network of canals and levées in the Atchafalaya basin. These had kept fresh water from some of the bayous, causing them to become stagnant. The floodwaters flushed them out and brought in fresh, oxygenated waters and nutrient-rich sediment.

In addition, the sediments carried down by the floodwaters helped build up new areas of marshland in the Basin. This was an important supply of sediment as the hard engineering strategies used to control the river over the decades since the 1927 flood have deprived coastal wetland areas of much needed silt. In addition, the flood sediment helped to flush out some of the oil that had covered the Louisiana coast following the devastating 2010 Gulf of Mexico oil spill.

Summary

- Significant flood events in the Mississippi, such as that in 2011, are triggered by physical causes including snow melt and heavier-than-usual rain. Once these physical factors trigger the heightened flood risk, the scale of the floods is increased by human factors such as levées and the tendency for development on the floodplain.

- The massive flooding of 2011 had mostly negative impacts on people, property and the land, resulting in 20 deaths, displacement of people from their homes, economic losses, damage to buildings and destruction of crops. However, there were some benefits to the land, including rejuvenation of marshland and helping to address the impacts of the 2010 oil spill in the Gulf of Mexico.

Topic 2 Ecosystems

The ecosystem as an open system

An ecosystem is a community of plants and animals interacting with each other (these are the organic or **biotic** components) and interacting with the environment in which they live (the inorganic or **abiotic** components). These abiotic components consist of **soils** and their characteristics and the **climate** along with other factors such as **relief**, **drainage** and **altitude**).

The term for all living plant and animal material in an ecosystem is **biomass**. We can use the **systems** approach to analyse ecosystems. Look for references to inputs, outputs, stores and transfers throughout this section.

Energy flows and trophic structures

The key concept here is that *energy flows through the ecosystem*. This means that it *enters*, *moves through* a series of trophic levels and then *leaves* via heat. Figure 14 outlines how this process occurs and introduces you to a range of terms that are used to describe elements of the process.

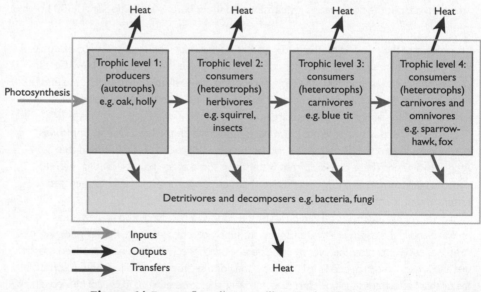

Figure 14 Energy flow diagram, illustrated with reference to Breen Wood, Ballycastle, Co. Antrim

Inputs of energy

Energy enters the ecosystem via photosynthesis, as green plants fix solar energy and convert it into chemical food energy.

Food chains and trophic levels

This input of food energy is the basis for a **food chain**, where the energy fixed by photosynthesis becomes available to subsequent levels of the ecosystem. Each level is known as a **trophic** (energy) **level**.

The first trophic level consists of the plants. As they can produce their own food directly, they are called **producers** (or **autotrophs**). All subsequent trophic levels have to consume plants and other organisms in the previous level, so they are called **consumers** (or **heterotrophs**).

In the second trophic level, the **herbivores** (plant eaters) consume plants from the first level. In turn, these herbivores provide food for the third trophic level, the smaller **carnivores** (meat eaters). The fourth, and usually final, trophic level is made up of the larger carnivores and **omnivores** (plant and meat eaters).

Thus food energy is transferred up through the trophic levels. This transfer of energy is not very efficient, however, as around 90% of the energy at one trophic level is lost and thus unavailable to the next trophic level. This energy is lost via life processes in organisms such as **respiration**, heat, excretion and non-consumed elements such as bones. Due to this inefficiency in energy transfer, there are usually *no more than four trophic levels* in an ecosystem.

At each trophic level, the **detritivores** (such as bacteria and fungi) are consumers that act to decompose dead organic matter.

The energy available at different trophic levels can be shown in a trophic pyramid (Figure 15).

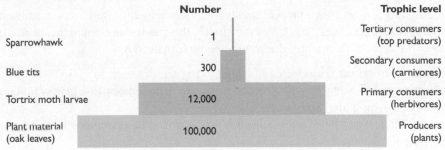

	Number		Trophic level
Sparrowhawk	1		Tertiary consumers (top predators)
Blue tits	300		Secondary consumers (carnivores)
Tortrix moth larvae	12,000		Primary consumers (herbivores)
Plant material (oak leaves)	100,000		Producers (plants)

Figure 15 Trophic pyramid for the deciduous forest at Breen Wood, Ballycastle

Nutrient cycles

The key point here is that *nutrients mainly cycle around between stores within the ecosystem* (although there are inputs and outputs as well). This is shown in Figure 16 — note the inputs, outputs, stores and transfers.

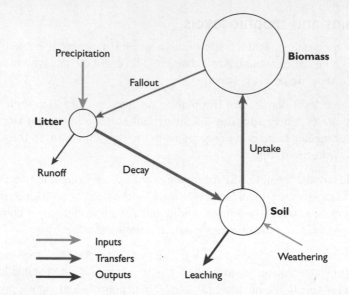

Figure 16 Gersmehl model applied to Breen Wood (deciduous forest)

The nutrients leave the **biomass store** and are transferred by **fallout** into the **litter store**. This dead organic matter is broken down by the decomposers and transferred by **decay** into the **soil store**. Here, the nutrients are available to plants via their roots and are transferred by **uptake** into the **biomass store** and thus the cycle continues. In addition, there are two **inputs** of nutrients (carbon dissolved in **precipitation** and minerals **weathered** from the bed rock) and two **outputs** (losses via **runoff** from the **litter store** and **leaching** from the **soil store**).

The diagram can be drawn proportionally to show the relative amounts of nutrients in the stores and transfers: the larger the circles, the greater amount in the stores; the thicker the arrows, the greater the volume being transferred.

The **rate of nutrient transfer** can be affected by factors such as:
- **climate** (in hot, wet tropical rainforests the rate of decomposition is very high and rates of leaching are considerable also)
- **soils** (acidic soils have fewer soil organisms and thus decomposition rates are slower)
- **vegetation type** (the leaves from coniferous trees take longer to decompose than those from deciduous trees)

Case study

Breen Wood

Abiotic components of Breen Wood deciduous woodland

Breen Wood deciduous forest near Ballycastle is one of the few remaining areas of the ancient deciduous woodlands that used to cover much of Ireland. Its **abiotic** components are:
- **climate** — average temperatures range from 4°C in winter to 15°C in summer and annual rainfall totals are quite high at around 1600 mm.
- **soils** — the **podsols** found here are generally poor in quality, partly due to the basalt parent rock, which is low in nutrients. Thus, the soils are thin and have an acidic pH of

around 4.5. Furthermore, the combination of high rainfall totals and sloping relief ensure that leaching is significant.

As a result of these abiotic factors, the range of plants and animals is smaller than would typically be expected in a deciduous forest, and the 200-year-old oak trees in Breen Wood are about half the size they would be in lower areas.

Energy flows in Breen Wood deciduous woodland

The first trophic level consists of the dominant producers in Breen Wood: *oak* and *birch* trees, forming a dense **canopy layer** 20m above the ground. Below these, a **shrub layer** is found consisting of *hazel* and *holly* trees. In the **ground layer**, plants such as *ferns*, *brambles* and *mosses* grow in the damp shade below the trees (Figure 17). All these autotrophs are able to fix solar energy via photosynthesis, making it available to subsequent trophic levels.

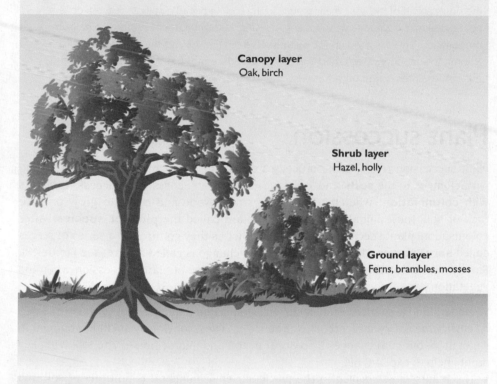

Canopy layer
Oak, birch

Shrub layer
Hazel, holly

Ground layer
Ferns, brambles, mosses

Figure 17 Layers of the deciduous forest at Breen Wood

The second trophic level consists of the herbivores, although the harsh climatic and soil conditions mean that the range of heterotrophs in the ecosystem is more limited than in other deciduous woodland areas. The herbivores in Breen Wood are mostly insects, including around 15 species of butterfly (*orange tip* and *speckled wood*), and *red squirrels*.

Some of this energy is available to the third trophic level, which consists of birds, including omnivores such as *blue tits* and *goldcrests*, which forage for seeds and insects. The fourth and final trophic level consists of carnivores and omnivores including *sparrow hawks* and *buzzards* (which feed on the smaller birds), *badgers* (which feed on earthworms, small birds, fruit and nuts), *foxes* (which eat the squirrels and smaller birds, as well as grass and fruit) and *stoats* (carnivores that feed on squirrels and birds).

At each trophic level, up to 90% of the available energy is lost, so the biomass of the fourth trophic level is considerably smaller than that of the first in Breen Wood. Also, at each level, detritivores such as *earthworms* are the consumers that help decompose dead organic matter. In Breen Wood, due to the acidic podsol soils, the work of the detritivores is limited and decomposition rates are slow.

Nutrient cycles in Breen Wood deciduous woodland

(Refer to Figure 16, p. 32 as you read this.)

Due to the acidic podsol soils and cool average summer temperatures of 15°C in Breen Wood, there are fewer soil organisms such as earthworms. This means that the rates of decomposition are quite slow, and it may take years for the leaves to fully break down and travel via the decay pathway to the soil store.

The high rainfall totals (around 1,600 mm) and cool summer temperatures, along with the sloping relief of the area, mean than leaching is quite high. The basalt bedrock does not release many nutrients via weathering, leaving the main nutrient input to come from precipitation. There is good uptake of nutrients from the soil into the biomass and each autumn the deciduous trees shed their leaves to reduce moisture loss via transpiration during the winter, contributing fresh leaf litter via the fallout pathway to the litter store.

Plant succession

Microclimate Small-scale climatic characteristics of an area that differ from the general climate. It can apply to an area of a few square metres to many square kilometres.

Primary succession is succession that occurs on previously unvegetated surfaces, for example sand dunes or bare rock.

Secondary succession occurs on previously vegetated surfaces that have lost their vegetation through processes such as fire or landslides.

Plant succession refers to the processes of change over time in the **plants** (along with changes to the **soils** and microclimate) in an ecosystem. Succession begins with **colonisation**, whereby the initial plants invade and begin to grow on bare rock or soil; these initial colonising plants are called the **pioneer species.** After colonisation, plant species replace one another as they go through a series of stages called **seral stages** (the whole sequence of change is called a **sere**, see Figure 18). Finally, **climatic climax** is reached — the final stage of plant succession when the vegetation has reached a balance with its environment.

To understand how succession occurs, it is important to realise that *various processes of environmental change occur* to produce it. These obviously include changes in vegetation, but vital changes also occur in soil and microclimate. The following key points help to explain this:

- The plants *most adapted* to the particular environmental conditions of soil and microclimate *dominate* at that given time (e.g. mosses colonise bare rock).
- However, these very plants tend to *modify the environment* in which they grow, improving the *soil conditions* by adding organic matter, increasing soil depth, aiding soil moisture retention and changing soil pH. They also improve the *microclimate* by reducing ground wind speed and sheltering the soil, thus reducing evaporation and maintaining higher soil temperatures.
- Thus they *create conditions more favourable for other plant types*, which will then come in and replace the previous plants (e.g. the moss will be replaced with taller plants such as grasses, which will in turn be replaced by shrubs). In addition, as the soil and microclimate become less harsh, a wider variety of plants can grow there, so the number of plant species tends to increase too, although each seral stage tends to have one dominant plant type.

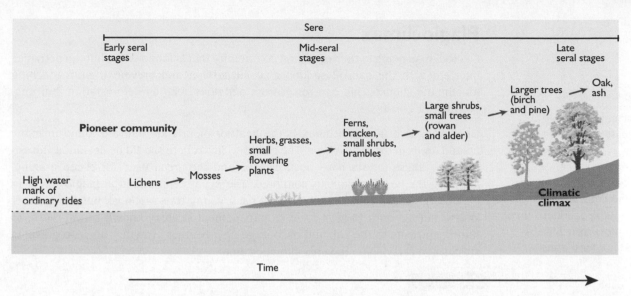

Figure 18 An example of plant succession

This means that change is self-perpetuating, as plants invade, dominate, change the environment and are replaced, until a climatic climax is reached and the plants reach a balance with the environment. Once the climatic climax vegetation is established, it tends to exclude its rivals and the number of plant species can drop slightly.

Table 3 Changes during succession

	Explanations for changes
Soil changes	
Depth increases	Due to the addition of more organic matter from the biomass and more weathering of the bedrock
Humus increases	More organic matter is added to the soil via nutrient cycling
Moisture increases	The humus aids moisture retention and the shade from plants reduces evaporation from the soil
Stability increases	The increasing root system binds the soil together
pH decreases	The soil becomes less alkaline due to the addition of humic acids
Colour darkens	More organic matter is added to the soil via nutrient cycling
Plant changes	
Biomass and plant longevity increase	The deeper and richer soil can support larger and more long-lived plants such as trees; plant abundance increases and the amount of land surface not covered by plants decreases
Species diversity increases	The improving soil and microclimatic conditions can support a wider variety of plant types, not just those adapted to the harsh conditions of colonisation
Stratification (different layers of vegetation)	The wider variety of plant types increases stratification
Microclimate changes	
Wind speed drops	The increased plant cover provides shelter and acts as a wind break

Examiner tip

It should be clear from these points that, while the *plants* certainly change during succession, there are also changes to the *soils* and the *microclimate*. Read the exam questions carefully to see which of these three you are being asked to discuss.

Knowledge check 10

State the **three** aspects of the environment in an ecosystem that can change as plant succession occurs.

Plagioclimax

Plagioclimax refers to the vegetation community that is found when human activities interfere with the natural sequence of succession and prevent it from reaching its climatic climax. This can result from activities such as deforestation, burning, draining or grazing.

One example of a plagioclimax is the heather moorlands that are found in many upland areas in the British Isles. The climatic climax here should be deciduous forest. However, these forests have been cleared and the areas used for sheep grazing, allowing the heather plant to dominate. The ecosystem is held at plagioclimax in part by the sheep grazing (which cuts back young tree saplings), but the heather is also burned on a 15-year cycle (a management strategy known as *muirburn*) to return nutrients to the soil and encourage new heather growth. This also prevents succession carrying on towards climatic climax.

Examiner tip
Include an example of a *cause* of plagioclimax (e.g. in heather moorlands) in your exam answers, expanding into some more detailed discussion of how one cause operates if there are more than 2 marks available for the answer.

Case study

A psammosere succession in the sand dunes at Portstewart Strand

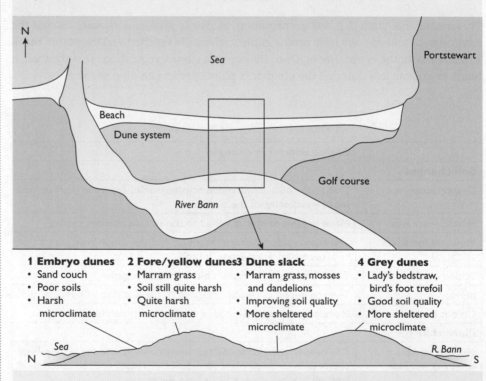

Figure 19 Succession in the psammosere at Portstewart Strand

The initial bare surface on which colonisation occurs is the sand at the very top of the beach. The conditions here are very tough: the *soil* has next to no organic matter, the salt levels are high, the pH is very alkaline due to the amount of sea shells in the sand and the *microclimate* is harsh, with no shelter from strong onshore winds blowing in off the Atlantic Ocean. Therefore, only very hardy plants adapted to these conditions can grow here. In Portstewart Strand this is mainly *sand couch*. This initial seral stage is referred to as **embryo dunes**.

As these plants colonise, they begin to change the environment, especially the soil. Their roots help to stabilise the soil and their leaves help the dunes to grow in height as they trap more sand. The soil slowly begins to improve as more organic matter is added when plants die. This begins to increase the humus levels, aiding water retention in the very sandy soils.

These improving conditions create the conditions more suited to other plants, especially *marram grass*. Marram grass begins to invade and dominate, forming the next seral stage — the **foredunes** or **yellow dunes**. Marram grass is particularly adapted to the still quite harsh conditions in the yellow dunes. Its leaves are tightly rolled and have a shiny surface to reduce moisture loss via transpiration. Its growth is stimulated by burial under sand and spreads readily via its rhizome root system. These roots further stabilise the soil and increase the dune height.

The sand dunes at Portstewart form a series of ridges roughly parallel with the beach. As you cross over the first ridge into the **dune slack**, the microclimate changes and becomes noticeably more sheltered. This, along with the improving soil quality, allows a wider variety of plant species to grow. At first, *mosses* grow in the spaces between the marram grass and rosette plants such as *dandelions* start to appear. These plants tend to spread laterally, covering the surface of the soil, reducing moisture loss via evaporation and helping the soil temperature to remain a little higher.

Further into the dune slack, these changing conditions allow plants such as *ribwort* to grow, and gradually the marram grass is replaced.

At the back of Portstewart Strand are the old, stable **grey dunes**. The pH here is between 6.5 and 7.5, and the soil has significant amounts of humus, supporting plants such as *lady's bedstraw* and *bird's foot trefoil*.

As this dune system is managed by the National Trust, the succession is prevented from reaching its climatic climax by strategies such as allowing cattle grazing, especially towards the western end, although the odd isolated *willow* or *hazel* tree can be found towards the back of the dunes.

Alistair Hamill

Portstewart Strand

Examiner tip

As in the previous case study, the *facts* for Portstewart Strand are mostly plant types. Make sure you know and include a wide range of plants to represent the various seral stages in the dune system.

Examiner tip

In longer exam questions on this case study, you will probably be asked to describe *and* explain the successional changes. Make sure you are able to clearly explain the processes of change in *soils* and *microclimate* brought about by changing *vegetation*.

Knowledge check 11

Making reference to your small-scale study of plant succession, describe **two** changes.

Human interaction with ecosystems in the North American prairies

Characteristics of the mid-latitude grasslands and the mollisol soils

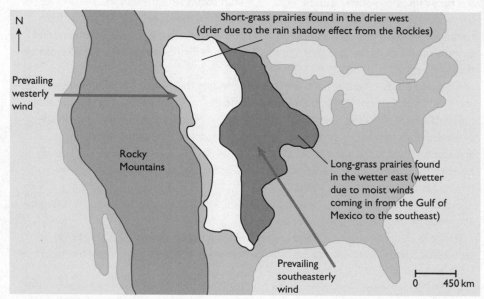

Figure 20 The location of the long-grass and short-grass prairies in North America

Climate

- **Temperature:** The prairies are characterised by *hot summers* (maximum temperatures can exceed 35°C) and very *cold winters* (when temperatures can drop below −30°C) and thus the *annual temperature range is very high*. The main reason for this is that they are far from the moderating influence of the sea and thus experience the effects of continentality.
- **Precipitation:** Overall, the levels of *precipitation* are *low*, with a typical range of 400–500 mm per year, as the prairies are far from the sea. Furthermore, even when rainfall totals are relatively higher in the spring and summer months, the warm temperatures mean that there is *significant evapotranspiration*. This further reduces the amount of water available in the soil. The rain can fall in intense thunderstorms, so much of it flows to rivers via overland flow without entering the soil.

There is noticeable *west/east variation in precipitation*, with *rainfall totals increasing as you move further east*. This is because the western prairies are under the influence of the rain shadow effect associated with the Rocky Mountains to the west. In contrast, the eastern prairies experience moist winds coming up from the Gulf of Mexico to the southeast.

Examiner tip
Some of the climatic factors mentioned here (including **continentality** and the **rain shadow effect**) are covered in the Atmosphere topic.

Vegetation

The climatic climax vegetation in this ecosystem in its natural state is around 50 different species of grass and so this is a **biodiverse** ecosystem. Additionally, these grasses are **stratified** into two main layers, with shorter, feather grasses such as *prairie dropseed* and *smooth brome* (which tend to have shallower roots that form a dense root web) mixed with taller, tufted grasses such as *prairie dock* and *big bluestem* (which tend to have longer, deeper roots).

The vegetation is adapted to the climate. Due to the low overall rainfall totals and the high rates of summer evapotranspiration, there is insufficient soil moisture to support trees. The dense coverage of grasses also restricts tree growth. During the autumn and before the onset of the cold, harsh winter, the grasses die back slightly to form a turf mat. During late spring and early summer, when temperatures rise and snow melts, the grasses grow rapidly. Their inward-curling blades are adapted to reduce transpiration in the hot summers and to retain as much moisture as possible.

The west/east climatic variations also affect the vegetation. In the wetter east, the prairie grasses are longer and include species such as *big bluestem*, which can be up to 2 m high. In the drier west they are shorter and include *buffalo grass* and *blue grama*.

Soils

The soils found in this ecosystem are known as **mollisols** (or **chernozems**) and are deep (around 1.5 m in depth), rich soils consisting of two main layers or horizons (Figure 21). The **A horizon** is a crumbly, black topsoil that is rich in **humus**. During the summer, the soil organisms such as worms help decompose the litter and incorporate it into this top layer, forming the rich humus.

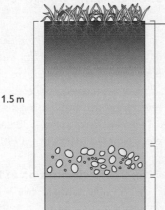

1.5 m

Thick, dense root mat

A horizon
Crumbly black top, rich in humus with crumb structure and neutral to slightly acid pH; low precipitation totals mean little leaching occurs

Cca horizon
Capillary action in summer leads to lumps of calcium carbonate in the Cca horizon

C horizon
Bedrock, rich in lime

Figure 21 Mollisol soil profile

> **Examiner tip**
> You might be asked to draw or complete an annotated soil profile for a mollisol, so learn this diagram well.

As precipitation only slightly exceeds levels of evapotranspiration the soils experience limited **leaching**, and so there is little loss of nutrients via this output. The leaching that does occur is mostly associated with the spring snowmelt and the heavy summer storms. In fact, the high summer temperatures and lack of soil moisture actually draw water *upwards* from the lower levels by **capillary action**. This leads to nodules of calcium carbonate in the upper C horizon (known as the **Cca horizon**). Consequently, the soil nutrient store is the largest in this ecosystem. In contrast, the biomass store is quite small due to the fact that the dominant plant species is grass).

Mollisols have a crumb structure and a neutral to slightly acid pH.

The overall result of these soil characteristics is that the mollisols are among the most naturally fertile soils in the world.

Case study

Human impacts and attempts to manage the grasslands: the North American prairies

Impacts of human activity

The main human activity in the prairies is **monoculture**, which is defined as the growing of the same crop year after year in an area. This is particularly common in the wetter east of the prairies. The main crop is cereals (such as corn and wheat). Monoculture has a number of negative impacts on *soils*:

- The repeated removal of vegetation takes nutrients from the ecosystem, disrupting the nutrient cycle.
- The ploughing of the soil breaks up the surface mull humus. This damages the soil's structure, making it more vulnerable to erosion. In addition, the loss of humus means the soil is less able to retain moisture.
- As part of the harvesting process, the soil is left bare and exposed to erosion by the wind and the intense rainstorms of the prairies.

One of the most infamous examples of soil erosion in the prairies was the 1930s Dust Bowl. Monoculture and other practices, such as the burning of crop stubble to control weeds, combined with a serious drought to cause extensive erosion. The soils were dried out and left bare allowing the strong winds to remove up to 75% of the topsoil in places. Some areas, such as the Badlands in Dakota, were left so severely eroded that farming is still next to impossible. And yet despite the grim experience of the Dust Bowl in the 1930s, around 9.5 million hectares of land were in fallow (and therefore exposed to erosion) in 1981, an increase of 42% since 1931.

The effects of monoculture have been exacerbated by the **mechanisation of farming** in the prairies. Not only has this allowed farming to occur on a much larger scale, but the tractors and other farm vehicles compact the soil, weakening its structure.

As a result of these problems, there has been a drop in farm yields in parts of the prairies. It was estimated that *Canadian Prairie farmers lost $100 million in 1984* as a result of soil degradation — a *10% drop in net farm income*.

Monoculture can also have a negative impact on the *vegetation* in the prairies:

- There is a significant reduction in biodiversity, with the range of plants such as *prairie dropseed* and *prairie dock* being replaced by a single cereal crop. A 1990 US government study suggested that 35 animal species and over 350 plant species were under threat.
- There is a reduction in stratification, with the multi-layered vegetation being replaced by plants of one height.
- The dense root system is broken up, leaving the soil more prone to erosion, and the plants with deeper roots (such as the *big bluestem*) are removed and replaced by cereals with shallower roots.

In the drier west, the main human activity is **livestock**. This can lead to problems such as **overgrazing**, which the US Environmental Protection Agency suggests is responsible for 28% of the soil erosion problem.

Evaluating the management of human impacts

A range of strategies have been used to try to reduce soil erosion in the prairies:

- **Mulching:** By using practices that conserve crop residue (i.e. leaving behind the stalks from which the crop has been harvested), enough ground cover can be maintained from when the fields have been harvested until the next crop is big enough to protect the soil from erosion. In the prairies, a minimum residue cover of 1,500 kg per hectare is needed to protect most soils from serious wind or water erosion.
- **Green manuring/cover crops:** Green manure crops can be used as an alternative to fallowing. Green manuring is the practice of growing a crop not for harvest, but to work back into the soil using tillage. Instead of resting the soil, the plant that is grown returns all the nutrients back to the soil. In the prairies, legume crops (peas, clovers, lentils etc.) are preferred because they add nitrogen fixed from the air to the soil. This practice means that the soil is not left bare for a whole year, reducing the wind and water erosion.
- **Shelter belts:** These are lines of trees planted at the edges of fields to act as wind breaks and shelter the soil. The Canadian Permanent Cover Program encourages farmers to enter into long-term contracts for 15 or 21 years. In Manitoba, the programme offers $40 per acre per year for 10-year and $70 per acre for 21-year agreements.
- **Contour ploughing:** This involves the ploughing of fields along the contour lines, thereby *not* creating ready made gullies that increase water erosion. This was first used in 1935 during the Dust Bowl when the Soil Erosion Service started to pay farmers to employ it. It is extensively used in the less flat areas of the prairies, such as Nebraska.
- **Strip farming:** This is the practice of growing crops in strips that alternate with strips of fallow. It reduces wind erosion by lowering the wind speed on the surface of the soil and the distance the wind travels across exposed fallow. In most areas of Saskatchewan, for example, the prevailing wind direction is northwest or west. Therefore the best direction for strips for reducing the erosion potential is north to south.

How effective have these strategies been?

These strategies have been effective to a certain extent. Soil erosion has been reduced: according to the 2007 US National Resources Inventory (NRI), between 1982 and 2007 soil erosion decreased by 43% and, although declines in soil erosion have moderated slightly since 1997, the trend is still downward. In the Southern Plains region soil erosion decreased from 12.5 tonnes per acre per year in 1982 to 8.8 tonnes per acre per year in 2007.

Examiner tip
When explaining these management strategies, make sure you link them directly and clearly to how they address the soil erosion problem.

Examiner tip
Look up annotated diagrams to explain these strategies on the internet and practise drawing them. Diagrams can aid the clarity of your explanations.

Examiner tip
Explain…or evaluate? Check the command phrase. If asked to *explain*, you should outline *how* the strategies operate. If asked to *evaluate*, you should make reasoned judgments about their effectiveness.

Despite this, in 2007 99 million acres of land were eroded above sustainable rates, according to the NRI report, resulting in the loss of 827 million tonnes of soil in these areas per year.

As a result, many are calling for a more sustainable approach to prairie management, which includes returning the prairies to their natural state. One such proposal is the Buffalo Commons — a suggestion to return an area of 360,000 km^2 to natural grassland conditions. Although initially controversial, the idea has been gaining popularity in recent years, with the Kansas City Star newspaper suggesting in 2009 that a 4,000 km^2 Buffalo Commons park be established in western Kansas.

Knowledge check 13

With reference to the prairies in North America, explain the impact of monoculture on the soils and vegetation of the ecosystem.

Summary

- An ecosystem is an open system consisting of abiotic components (climate and soils) and biotic components (plants and animals) interacting with each other.

- After entering the ecosystem via photosynthesis, energy flows up a food chain/web through a series of trophic levels. The energy transfers are very inefficient, however, and much of the energy is lost before it reaches the next trophic level. As a result, ecosystems seldom have more than four trophic levels.

- Nutrients cycle around between three stores within the ecosystem (biomass, litter, soil) through a series of transfers or pathways (fallout, decay and uptake). In addition, there are inputs (via precipitation and weathering) and outputs (via runoff and leaching) of nutrients from the ecosystem.

- Plant succession refers to the process of change in vegetation over time whereby one dominant plant species replaces another dominant species until the plants reach a balance with the environment known as the **climatic climax**.

- The initial **pioneer plants** must be adapted to harsh conditions of soils and microclimate. But they begin the process of improving the soil and microclimate, making the conditions more favourable for other plants. Thus vegetation coverage, plant height and species diversity increase as succession continues.

- If humans stop the natural process of succession by activities such as grazing or burning, the stage at which it is held is known as a **plagioclimax**.

- The prairies are characterised by low levels of precipitation and a high annual temperature range. The climatic climax vegetation is grass — long grass dominates in the wetter east of the prairies and short grass in the drier west. The grasses are stratified and biodiverse. The soils are mollisols and consist mostly of an A horizon that is rich in humus and a Cca horizon. Leaching is limited and during the summer soil moisture moves upwards via capillary action.

- The main human activity in the prairies is monoculture. This has a number of negative impacts on the soil, especially erosion, which was particularly bad during the Dust Bowl in the 1930s. There are also negative impacts on the vegetation as biodiversity and stratification are reduced.

- These human impacts are managed in various ways including mulching, green manuring, shelter belts, contour ploughing and strip farming. These have been successful to a certain extent. However, soil erosion is still occurring and so others are calling for a more sustainable management of the prairies, including the Buffalo Commons concept.

Topic 3 Atmosphere

Atmospheric processes

The global energy balance

The Earth's atmosphere extends about 1,000 km from the surface of the Earth but the majority of the gases needed to support life are found within the lower 40 km of the atmosphere. Most of the weather and climate within the atmosphere takes place within the troposphere (which is about 16–17 km deep).

The global energy balance is an open system where the Earth receives heat/solar energy (insolation) and energy is stored and transferred continuously between the Earth and its atmosphere. Some energy will eventually be reflected back into space. The amount of energy reflected depends on the Earth's albedo.

How is heat transferred around the world?

Vertical heat transfer

Heat can be transferred from the surface of the Earth to the atmosphere, having a cooling effect on the Earth while warming the air. This can happen in one of three ways:

- **Radiation:** The land radiates heat back out to space through long-wave radiation.
- **Convection:** Warm air is forced to rise as part of convection currents. The rising warm air is replaced by colder, descending air.
- **Conduction:** Energy is transferred through contact.

Horizontal heat transfer

Heat across the world is not distributed equally. Between the equator and 40° N and S there is a heat surplus while from the poles to 40° N and S there is a heat deficit. The atmosphere redresses this in the following ways:

- **Ocean currents:** The North Atlantic Drift, for example, brings warm water from the Caribbean towards Europe while the Labrador current brings cold water down from the Arctic to the east of Canada (especially in the winter).
- **Winds:** Winds carry surplus energy from the tropic areas away from the equator. As the warm wind blows over the sea (or vegetation) it picks up moisture and in the process stores latent energy.
- **Hurricanes:** A major weather system can sometimes occur that involves ocean currents, warm seas and tropical winds, which lock latent heat in and then transfer it gradually across the globe.

The global temperature pattern

Latitude is the most important factor when looking at the pattern of global temperature. Areas at the equator usually receive 12 hours of sunlight. The angle of the sun concentrates any rays at the equator, but the angle becomes more acute

Insolation Radiation/energy that comes from the Sun and enters the Earth's atmosphere.

Albedo The amount of reflection that occurs when solar energy meets clouds, dust or the surface of the Earth. It is shown as a percentage of all of the incoming radiation.

Examiner tip
Make sure that you understand the key terms and their definitions.

Knowledge check 14
Describe the **three** processes that can transfer heat vertically in the atmosphere.

Latent heat As water changes state from liquid to gas (evaporation), some energy is stored due to the process.

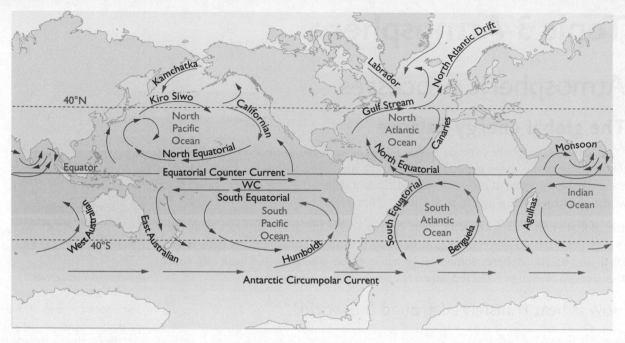

Figure 22 Ocean currents

towards the poles, with less energy being spread over a much larger area. However, variations are obvious within this pattern as other, more localised factors have their influence:

- **Continentality:** Areas that are much further inland experience bigger temperature ranges and extremes than areas on the coast. For example, the UK average temperature is between 0°C and 8°C, whereas Siberia on a similar latitude has average temperatures of between –20°C and –10°C.
- **Ocean currents:** The sea absorbs heat at a much slower rate than the land. Ocean currents bring warmer water to some areas and colder water to others (Figure 22).
- **Altitude:** Large expanses of high mountain ranges lose around 0.6°C for every 100 m of height.
- **Prevailing winds:** These cause some localised areas to be warmer/drier than others.

The factors that control wind speed and direction

What is wind?

Pressure is the weight of the air at a point within the atmosphere. Wind is the movement of air that results from the atmosphere redressing pressure differences.

What controls wind direction?

There are three main factors that influence the direction in which wind blows:

- **Pressure gradient:** Air flows from high to low pressure along a gradient (the difference in the pressure divided by the distance). For example, the trade winds blow from areas of high pressure to areas of low pressure.
- **The Coriolis effect:** The rotation of the Earth deflects the flow of air in the northern hemisphere to the right (and in the southern hemisphere to the left). The further north a place is, the more impact that the Coriolis effect has — areas 50–55°N have winds that are almost at right angles to the pressure gradient.
- **Friction:** Friction between the air and the surface can limit the impact of the Coriolis effect.

The resultant impact of all three controls is that winds do not travel directly from high to low pressure but spiral outwards from the centre of high-pressure areas towards areas of low pressure.

What controls wind speed?

There are four main controlling factors that influence the speed of the wind:
- **Pressure gradient:** The bigger the pressure gradient (i.e. the difference between the high and low pressure) then the more powerful the wind. Pressure is shown on weather maps as isobars. The closer the isobars, the stronger the wind.
- **Friction:** Wind can be slowed down by contact with other surfaces (like water and the land).
- **Turbulence:** Internal friction (within the air) is caused by uneven surfaces.
- **Local factors:** Sometimes buildings and other local features can influence wind patterns, and either slow down the wind or create a 'wind tunnel' effect.

The general circulation of the atmosphere

The general circulation of the Earth's atmosphere helps to explain the pattern of wind and pressure belts. This circulation system is the main way in which the Earth can redistribute energy received at the equator towards the poles. It is controlled by three cells (Figure 23).

1 Development of the Hadley cell

The Earth's atmospheric engine gets its energy from the direct solar heating at the equator. Heat is transferred to the air above and the air rises and cools. Rising air means that low pressure is left at the surface, known as the ITCZ (inter-tropical convergence zone).

The rising air diverges and flows towards the poles (both north and south) and sinks again around 30° N and S of the equator. This warms and produces high pressure. The air is then diverted back towards the equator at a much lower level while some air escapes into the mid-latitudes.

2 Development of the polar cell

The extreme cold air at the poles leads to subsiding air (and high pressure at the surface). The air then moves away from the high pressures across the surface towards the 60° N (and S) point where the air begins to rise at what is called the polar front.

Knowledge check 16

How does the Coriolis effect work differently in the northern and southern hemispheres?

Examiner tip

Make sure that you understand and can explain fully the difference between the controls on wind direction and wind speed. Exam questions will often try to get you to consider the different ways in which these operate.

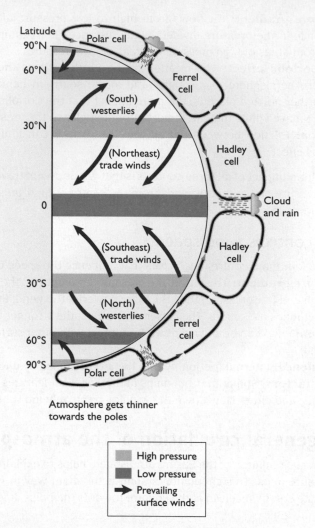

Figure 23 Tri-cellular model of atmospheric circulation

3 Development of the Ferrel cell

Some of the air from the Hadley cell moves across the surface as westerlies. At 60° N (and S) a low pressure forms as warm air from the equator meets blocks of cold air from the poles.

Moving from the equator to the poles there are alternating bands of high and low pressure. However, in some cases this pattern is modified by other factors, including continentality. For example, Siberia should be located in a band of low pressure throughout the year, but it experiences high pressure in the winter as a result of the extreme cold of the continental interior causing the air to subside.

The distinction between absolute and relative humidity

The amount of water vapour in the air is called humidity.

Knowledge check 17

Describe how heat energy is circulated from the equator to the poles using the tri-cellular model.

Examiner tip

You need to know the different air movements associated with each part of the tri-cellular model at both the upper atmosphere and surface levels. Practise drawing the diagram for each cell.

Absolute humidity is the actual amount of water vapour per unit volume of air (measured in grams per metre cubed or gm^{-3}). Warm air can hold more water vapour than cold air.

Relative humidity (RH) is the amount of water vapour in the air expressed as a percentage of the total water vapour air can hold at a particular temperature. If the RH of the air reaches 100%, it is called saturated, and the dew point is reached. RH is controlled by temperature, so that if air is heated, its capacity for holding more water vapour also increases.

Examiner tip

It is essential to have a good, detailed understanding of how relative humidity works.

Dew point temperature and the various causes of precipitation

The **dew point** is the temperature at which a parcel of unsaturated air becomes saturated, to achieve a relative humidity of 100%. The water vapour in the air then begins to condense, clouds form and eventually water droplets become too heavy for the air and fall towards the ground surface as precipitation.

Precipitation Any form of water moving through the air towards the surface of the Earth. It includes drizzle, rain, fog, mist, dew, hoar frost, hail, sleet and snow.

What are the main causes of precipitation?

Orographic (relief) rainfall

As a parcel of air is moved (by wind) from the sea towards the land it often is forced upwards as the hills or mountains cause a physical barrier. As the air is forced to rise, it expands, cools adiabatically (where there is no transfer of heat) and the relative humidity increases. At 100% RH the dew point is reached, resulting in precipitation.

Convectional rainfall

Convection currents take place when the air above the surface of the Earth is heated by the direct heating of the sun. The air is forced to expand, it cools adiabatically, and relative humidity increases. At 100% RH the dew point is reached and the air condenses, producing towering cumulonimbus clouds. These create unstable air, which causes thunderstorms and in some cases the conditions for a hurricane to form. Rain events can be short but intense.

Knowledge check 18

For each of the three main causes of precipitation, use an annotated diagram to help explain how they allow precipitation to develop.

Frontal (cyclonic) rainfall

Frontal rain occurs when two bodies of air (air masses) meet. This can happen at the polar front when warm air from the south meets with cold air from the north. At a front the air can become very unstable, because the warm and cold air cannot mix. The warm air starts to rise up and over the colder (dense) air. As the air expands, it cools adiabatically, leading to precipitation. Frontal rain is often associated with the development of depressions.

Examiner tip

Make sure that you have a detailed understanding of how each of the causes of precipitation works. They are all based on air being forced to rise.

Summary

- The global energy balance is an open system with insolation, storage of heat and some heat lost back into space.
- The global temperature pattern is affected mostly by latitude, but also by ocean currents, continentality, altitude and prevailing winds.
- Surplus heat is transferred around the globe horizontally (through ocean currents, winds and hurricanes) and vertically (through radiation, convection and conduction).
- Wind occurs due to differences of pressure in an area (the pressure gradient). The bigger the difference in pressure, the stronger the wind will be.
- Wind direction is affected by three controlling factors: pressure gradient, the Coriolis effect and differences in relief.
- Wind speed is influenced by four different factors: pressure gradient, friction, turbulence and local factors.
- The general circulation of the atmosphere is dominated by the tri-cellular model. This helps to explain how the Earth can redistribute energy received at the equator towards the poles.
- The dew point is the temperature to which a parcel of unsaturated air must be cooled so that it becomes saturated (an RH of 100%) in preparation for precipitation.
- There are three main types of precipitation: orographic, convectional and frontal (cyclonic).

Mid-latitude weather systems

Mid-latitude frontal depressions: their structure, formation and associated air masses

The British Isles is a good example of a geographic area within the mid-latitudes, where there is a mixing of air between the Ferrel cell and the polar cell (at the polar front).

What are air masses?

An air mass is a large parcel of air that can be thousands of kilometres wide. It remains in one location for a long period of time and picks up the area's temperature and moisture characteristics. The air mass can then move steadily across an area, bringing constant temperature and humidity conditions.

The five air masses that affect the British Isles (Figure 24) are:
- tropical maritime — warm and moist
- tropical continental — warm and dry
- polar maritime — cold and (fairly) moist
- polar continental — cold and dry
- arctic — cold, but with snow

Knowledge check 19

Describe the key characteristics of the dominant air mass that affects the British Isles.

Examiner tip

You need to know the subtle differences between each of the different air masses.

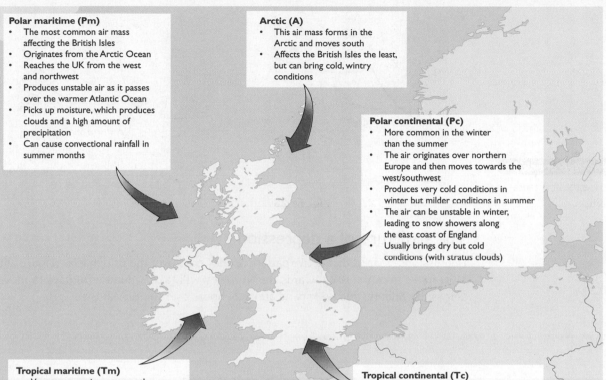

Polar maritime (Pm)
- The most common air mass affecting the British Isles
- Originates from the Arctic Ocean
- Reaches the UK from the west and northwest
- Produces unstable air as it passes over the warmer Atlantic Ocean
- Picks up moisture, which produces clouds and a high amount of precipitation
- Can cause convectional rainfall in summer months

Arctic (A)
- This air mass forms in the Arctic and moves south
- Affects the British Isles the least, but can bring cold, wintry conditions

Polar continental (Pc)
- More common in the winter than the summer
- The air originates over northern Europe and then moves towards the west/southwest
- Produces very cold conditions in winter but milder conditions in summer
- The air can be unstable in winter, leading to snow showers along the east coast of England
- Usually brings dry but cold conditions (with stratus clouds)

Tropical maritime (Tm)
- Very common air mass over the British Isles
- Air travels from the warm southern Atlantic Ocean (Azores), arriving from the southwest
- Warm air and sea temperatures allow the air to hold a high amount of water vapour, bringing a lot of precipitation to the British Isles
- Brings mild conditions in winter and warm, wet conditions in summer
- Causes dull skies (nimbostratus clouds), drizzle and fog (poor visibility)

Tropical continental (Tc)
- This air mass occurs rarely but usually occurs in summer
- Air travels from north Africa (the Sahara high-pressure zone) and across the Mediterranean
- Brings very warm and dry air from the south and southeast
- Brings milder conditions in winter and can introduce hot (heat wave) conditions in summer
- Can cause thunderstorms to develop if the temperature rises

Figure 24 Air masses affecting the British Isles

The formation of a depression

Depressions are areas of low atmospheric pressure that can produce cloudy, rainy and windy weather conditions. The depressions that affect the British Isles are formed out in the Atlantic Ocean when cold polar maritime air from the north moves south and meets warm, tropical maritime air that is moving north. The lighter, warm air will start to rise up and over the colder and denser air from the north. The rising warm air means that atmospheric pressure is reduced (leaving a low pressure). This leads to a disturbance called a baroclinic instability, which continues to develop into a front (Figure 25).

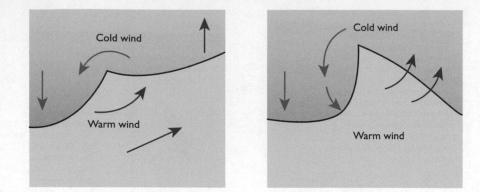

Figure 25 Formation of a depression

The structure of a depression

Depressions create a distinctive pattern as they pass over the British Isles (Figure 26). Most move from the SW towards the NE. They will take a number of days to move across the Atlantic and will spend about 24 hours over the British Isles.

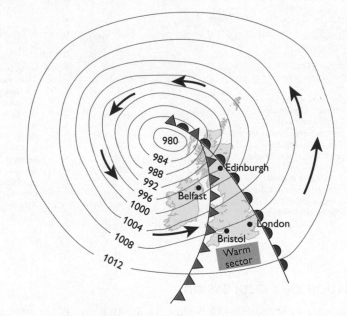

Figure 26 Depression over the British Isles

As the cold air gives way to the warm air from the southwest (the warm sector), changes to the weather result. During a depression the pressure is typically low (below 1000 mb) and falling. Depressions often encourage wind speeds to increase — isobars on weather maps will be close together. Temperatures vary depending on the type of air passing overhead and the time of year — they are lower in the winter months.

The depression creates a wide range of clouds and corresponding precipitation that varies as the depression passes overhead (Figure 27).

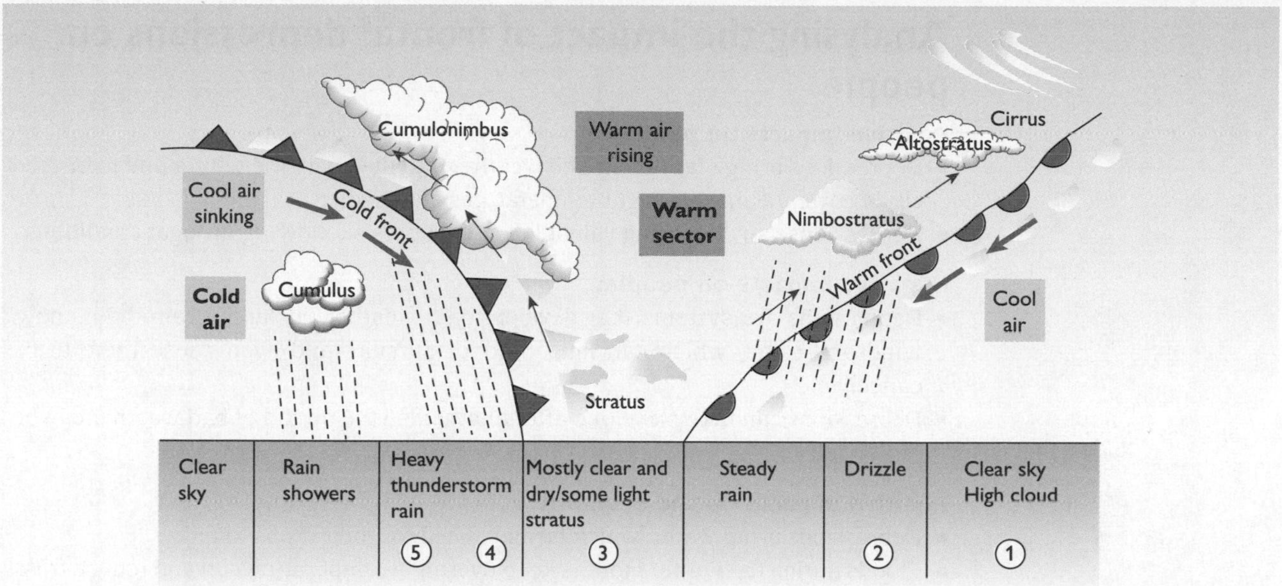

	⑤ Cold sector	④ Passage of cold front	③ Warm sector	② Passage of warm front	① Ahead of warm front
Pressure	Rising	Starts to rise	Steadies	Low	Starts to fall steadily
Temperature	Cold temperatures	Temperatures start to drop quickly	Warm	Temperatures rise	Quite cold
Cloud cover	As cumulonimbus clouds pass they give way to 'fair weather' cumulus clouds	Clouds thicken quickly with some towering cumulonimbus	Some clear skies with some light stratus clouds	Clouds are lower and thicken (altostratus and nimbostratus)	Cloud base drops and clouds get thicker (cirrus and altostratus)
Wind speed and direction	Wind speeds again start to decrease as depression passes (NW)	Wind speeds increase, sometimes to gale force (SW to NW)	Calmer conditions (SW)	Strong winds (SE to SW)	Wind speed increases and wind direction changes (SE)
Precipitation	Some rain showers, which gradually ease	Heavy thunderstorms with a chance of thunder and lightning; heavy rain and even hail	Rain, turning to drizzle and then dry conditions for a short period	Some sustained drizzle due to nimbostratus clouds	Little at first

Figure 27 Passage of a depression

Examiner tip

You might be asked to show detailed understanding of the formation and structure of a depression, so make sure that you understand this fully. Questions often ask you to draw a diagram of the passage of a depression.

Knowledge check 21

Describe and explain the sequence of events that are associated with a typical depression passing overhead.

Analysing the impact of frontal depressions on people

Positive impacts on people:
- Depressions bring warmer air and weather into the UK in the winter and increased cloud cover means overnight temperatures stay higher.
- Depressions can also bring valuable water supplies following drought conditions.

Negative impacts on people:
- Depressions are systems that develop 'bad' weather, including rain, hail, snow, wind and cloud, which can limit outdoor pursuits and even cause them to be cancelled.
- Depressions can take place one after another, so the period of bad weather can be prolonged.

Positive impacts on the economy:
- Depressions bring water, which farmers need for their crops to grow.
- Clouds during the winter months keep overnight temperatures up and reduce frost, prolonging the growing season.

Negative impacts on the economy:
- High wind speeds during the passage of a depression can damage crops and property, and cause the cancellation of flights and ferries.
- Flooding resulting from high amounts of rainfall can damage property.

Case study

The effects of low-pressure systems: the impact of the February 1994 storm on Northern Ireland

- The depression centre moved across Ireland on 2–4 February 1994.
- The pressure on 2 February was as low as 954 mb, and continued to fall to 950 mb on 3 February.
- An occluded front passed across Northern Ireland on 2 February, bringing a single band of cloud that produced 25 mm of rain across County Down and Antrim.
- On 3 February gale force winds up to 90 mph battered Northern Ireland.
- Two people died in the storm — one farmer was electrocuted by cables and another fell to his death during a power cut in County Antrim.
- Northern Ireland Electricity (NIE) recorded the largest failure of electricity services in over 50 years. 400,000 homes were left without any supply.
- Travel links were disrupted due to falling trees, and ferry services and air travel were affected on 3 and 4 February.
- Environmentally, flooding occurred on farmland near Newry and also around Dublin, where coastal defences were undermined.

Examiner tip
Make sure that you understand the impacts of this storm — both positive and negative — on the people and economy of Northern Ireland.

Understanding the formation of anticyclones and their associated weather

The formation of anticyclones

Anticyclones (Figure 28) are areas of high atmospheric pressure, which can produce calm, settled weather with little cloud cover or precipitation. Temperatures in the summer can be quite high and are associated with 'good' weather.

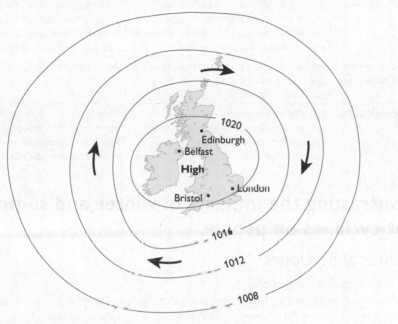

Figure 28 Typical synoptic chart for an anticyclone

The high-pressure system brought in by an anticyclone usually comes from the south of the UK. Tropical continental air masses arrive from areas of high pressure and bring warm and dry air into the British Isles. As the air sinks from a high altitude, condensation cannot take place. Anticyclonic conditions can also occur when the polar continental air mass develops high pressure and sinking air conditions.

Anticyclones can move into an area for a few days or sometimes for a longer period of time, blocking out any other weather system.

Weather associated with anticyclones

The weather produced by an anticyclone working in the summer can be very different from what happens in the winter (Table 4).

Table 4 Weather associated with anticyclones

	Summer anticyclone	Winter anticyclone
Pressure	High and increasing (over 1,000 mb)	High and increasing (over 1,000 mb)
Temperature	Temperatures can be warm — air from the warm south allows temperatures on hot sunny days to go above 24°C	Winter temperatures are much lower — the sun is low in the sky and there is less heat
Cloud cover	Sinking air allows for very settled/stable conditions and during the day temperatures increase; however, clear skies can lead to cold nights	As in summer, but lack of cloud cover at night can cause temperatures to go below freezing and lead to frost and ice
Wind speed and direction	Isobars are far apart — conditions are calm and any wind moves in a clockwise direction	As in summer
Precipitation	Very little, if any, precipitation, but sometimes dew or fog in the evening; thunderstorms can develop if heat leads to convectional rainfall conditions	Very little, if any, precipitation but sometimes freezing fog and frost/icy conditions can be caused by the lack of cloud cover

Knowledge check 22

Describe the main differences and similarities between winter and summer anticyclones.

Examiner tip

Make sure that you know and understand the main differences in how an anticyclone might operate in the summer and the winter. The differences are subtle but they can trigger very different weather features.

Contrasting the impacts of winter and summer anticyclones on people

Winter anticyclones

Positive impacts on people:
- Winter anticyclones can bring nice days during the winter with no cloud, no rain and a break from depression conditions.
- Although they can bring very cold temperatures and frosts, these can help to remove disease-causing organisms that might be flourishing within the soil.

Negative impacts on people:
- People with respiratory illnesses like asthma find it increasing difficult to breathe in cold weather.
- Frost and ice can be a hazard for drivers and for older people who might slip and fall.

Positive impacts on the economy:
- None really!

Negative impacts on the economy:
- Fog caused by temperature inversions can restrict transport and close airports.
- Ice and wintery conditions on roads can cause an increase in road traffic accidents, which can slow the movement of goods.
- People need to use more heating during cold conditions, which costs more money.

Summer anticyclones

Positive impacts on people:
- Anticyclones can bring 'good' weather, which lifts people's spirits and encourages them to be outside and enjoy more activities.

Negative impacts on people:

- Anticylones can bring long periods of hot and dry weather, increasing the risk of drought, which can lead to hosepipe bans and soil erosion.
- There can be a negative health impact as many old people and those with underlying medical conditions can be affected in dry, hot spells, and there is an increased risk of heat exhaustion.

Positive impacts on the economy:

- Anticyclonic weather, if not too prolonged, can be ideal for crops, allowing them to ripen fully.
- Tourism can be boosted by good weather. People take trips closer to home and spend money in the local economy.

Negative impacts on the economy:

- Anticyclones can bring drought, which can ruin crops and cause farmers to lose money.
- Irrigation might be used to artificially water the land if rainfall is unreliable, but this can be very expensive and do more damage than good if not managed carefully.

Case study

The effects of high-pressure systems: the impact of the 1976 summer anticyclone on people in England

- The 1976 heat wave led to the hottest summer average temperatures in the UK since records began, and the whole country suffered a severe drought.
- For 15 consecutive days temperatures stayed above 30°C.
- Parts of the country went for 45 days without rain through July and August.
- There were a number of heath and forest fires, and crops failed due to the drought, causing over £500 million of damage. Food prices increased by nearly 15%.
- Reservoirs and water supplies were at an all-time low, and some rivers ran completely dry. Some towns in Yorkshire had a total water restriction throughout most of the day. The government had to pass the Drought Act to forbid water usage for non-essential uses.
- Longer-term issues were noted in the years that followed — foundations that had used clay were found to have dried out and crumbled, causing houses to subside.

Examiner tip
Make sure that you develop detailed knowledge and understanding of the different impacts that this anticyclone had on the people and the economy of England. While anticyclones can bring nice, settled weather in the short term, the long-term consequences of a 'good spell' of weather can be disastrous.

Summary

- An air mass is a large parcel of air that stays in one location and picks up the area's temperature and moisture characteristics.
- The five air masses that affect the British Isles are:
  - tropical maritime (Tm) — warm and moist
  - tropical continental (Tc) — warm and dry
  - polar maritime (Pm) — cold and (fairly) moist
  - polar continental (Pc) — cold and dry
  - arctic (A) — cold but with snow
- Depressions are formed when polar maritime air from the north moves south and meets warm, tropical maritime air that is moving north. This causes a baroclinic instability, which creates a front.

- As a low-pressure system passes overhead the weather changes. The different weather elements are affected in different ways as the cold air, warm sector and cold air pass.
- Anticyclones are areas of high atmospheric pressure that produce calm, settled weather with little cloud or precipitation.
- Anticyclones can move into an area for a few days or they might remain for a longer period of time and block any other weather system.
- Anticyclones can cause slightly different weather features in summer and winter.

Extreme weather events

The formation and structure of hurricanes

Hurricanes are sometimes known as tropical revolving storms. They are severe tropical depressions where the wind speeds go above 64 knots (114 km h^{-1}). They happen when the pressure is extremely low (below 970 mb), sometimes as low as 880 mb.

The formation of hurricanes

Hurricanes happen most often in the late summer and autumn. They always track from east to west and always spin away from the equator. The actual mechanics of how tropical cyclones start to form is still not entirely understood by scientists, but the following six factors are thought to be necessary:

- Hurricanes form within the tropics, but not usually within 5° N or S of the equator (this is because the Coriolis effect here is not strong enough to get the hurricane to start spinning).
- Seawater temperatures need to be above 27°C (to a depth of 50 m) so that a lot of water can be evaporated quickly. This also allows enough heat to be in the air to allow the air to rise, causing convection currents and leading to condensation and cloud formation.
- Rapid cooling takes places as the air rises into high levels within the atmosphere. Clouds form into spiral bands.
- High relative humidity is needed so that a massive amount of moisture can be dealt with efficiently (over 200 million tonnes of water a day can be recycled by a hurricane).
- Low wind speeds are needed initially, but as the process of hurricane formation continues the wind speed within the hurricane can reach over 120 km h^{-1}.
- In order to form, a hurricane needs some sort of disturbed weather/depression.

The structure of hurricanes

Hurricanes can be massive weather systems spanning over 1,000 km and reaching up to 12 km into the troposphere/atmosphere. The structure of a hurricane is shown in Figure 29.

Hurricanes only start to lose power, intensity and wind speed when they lose their energy supply (i.e. warm sea water). This means that hurricanes fade when they pass over colder water or hit landfall.

Knowledge check 23

Describe and explain some of the key factors required in the formation of a hurricane.

Examiner tip

You need to know the difference between the conditions needed for a hurricane to form and the structure and features of a hurricane when it is fully formed.

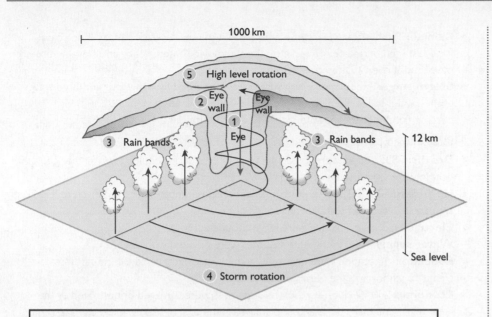

1 **The eye:** air here is subsiding, which gives a period of relative calm

2 **The eye wall:** air here is rising quickly due to strong convection currents, and surface air is drawn upwards quickly. Cumulonimbus clouds form and stretch up to the top of the troposphere (12 km)

3 **Rain bands:** a series of rain bands forms on either side of the eye walls as there are further movements of air upwards due to convection currents

4 **Storm rotation:** In the northern hemisphere the storm rotates, with surface winds spiralling in towards the eye (in an anticlockwise direction)

5 **High-level rotation:** at higher levels the air moves in the opposite direction (clockwise in the northern hemisphere)

Figure 29 Structure of a hurricane

Analysing the effects of hurricanes on people and property

Case study

Hurricane Katrina (USA), 29 August 2005

Hurricane Katrina was a category 5 hurricane. It started to form in late August in the Atlantic Ocean. The hurricane tracked to the west and moved across the Bahamas and through Florida on 25 August. It continued to generate energy in the warm seas of the Gulf of Mexico and moved north, hitting New Orleans on 29 August. On 28 August the mayor of New Orleans, Ray Nagin, when announcing the first ever evacuation of the city, noted that Katrina was 'a storm that most of us have long feared'.

What were the main hazards?

- **High wind speeds:** As a category 5 hurricane, Katrina sustained wind speeds of over 250 km h^{-1}. At its height, winds of 282 km h^{-1} were recorded.

> **Examiner tip**
> Practise drawing and labelling the cross-section of a hurricane and ensure that you can explain the different processes in detail.

- **Rainfall:** The hurricane brought a large amount of rainfall, which battered the coast for a sustained period before and after the storm had passed. Peak rainfall levels were measured at over 250mm over a 24-hour period.
- **Storm surge:** An 8.5-metre-high surge of water caused by high winds and large waves pushed water inland, flooding the New Orleans flood levée system. 80% of the city became flooded and floodwaters were pushed 19km inland.

The effect on people

- **Deaths:** 1,836 people were killed (mostly poor or elderly people), but a further 135 people were still classed as missing in 2013.
- **Homeless:** 500,000 people were listed as homeless. People found it difficult to leave the area and had to suffer terrible living conditions.
- **Electricity:** 3 million people were left without electricity.
- **Water supply:** All drinking water supplies were polluted with sewage.
- **Education:** 18 schools were destroyed and 74 were badly damaged, causing long-term effects to schooling in the New Orleans area.
- **Economic:** 230,000 jobs were lost as businesses were damaged or destroyed by the hurricane event. Many ports and oil platforms in the Gulf of Mexico were damaged and had to close. Damage was estimated at $108 billion — the costliest hurricane in US history.
- **Health:** Hazards included contamination of water by sewage, refuse and dead bodies.
- **Law and order:** The city suffered huge amounts of looting and other crime in the aftermath of the hurricane, and the US military was called in to restore law and order in some areas.

The effect on property

- **Homeless:** An estimated 300,000 houses were destroyed by the hurricane event, with around 500,000 people listed as homeless. Katrina redistributed over 1 million people across the USA. Over 100,000 temporary houses were built — many of them in Houston, Texas. The population of the state decreased by 8%.
- **Storm debris:** In some places within New Orleans, storm debris remained until 2010. Many damaged homes are still lying vacant and requiring serious rebuilding work 9 years later.
- **Rebuilding:** 6 months after the hurricane, the city centre in New Orleans was still without a working sewerage system, or gas and electricity supply.

Examiner tip

Any question that asks about the effect of a hurricane will require you to have detailed knowledge of the impact on both the people and the property in the area. Make sure that you have command of your case study.

Evaluating the protective measures used to reduce loss of life and damage to property

Case study

Hurricane Katrina (USA), 29 August 2005

Positive measures

Early warning systems: The USA has a very organised system for monitoring and predicting the track of hurricanes and major storms called the National Hurricane Centre (NHC). It was able to give accurate forecasts about the track of the hurricane and warn where the major impacts might be felt.

Evacuation plan: New Orleans had a comprehensive disaster and evacuation plan in readiness for an event like Katrina. The issue was that many people chose to ignore the warnings.

Emergency service preparation: The government authorities had practised their responses to a storm event like this. They created a fictional hurricane called 'Hurricane Pam' in 2004 and modelled the different responses needed in order to cope with a category 3 hurricane. One failure of this exercise was not to consider the impact of levée collapse, and yet it estimated that 60,000 people would be killed.

Negative measures

FEMA and disaster planning: The Federal Emergency Management Agency (or FEMA) began to assist the local authorities before the storm and continued for more than 6 months after the storm. A year after the storm, the former FEMA chief noted that 'there was no plan' for dealing with the aftermath of Katrina.

Evacuation planning: 25% of the New Orleans population did not own a car, so leaving the city was difficult. There were huge tailbacks as people tried to flee the city. The mayor, Ray Nagin, was criticised for delaying his emergency evacuation order until 19 hours before landfall.

Stranded people: Shortly after the hurricane had taken place, media images began to appear of people who were left stranded by floodwaters without food, water or shelter as death rates started to mount up. Many felt that the government response was slow and delayed in getting the right aid to the right people.

Emergency service preparation: The breaching of the levées meant that mass flooding became a huge issue in New Orleans. Aid could not enter the areas of need using overland routes. The Louisiana Superdome was the 'shelter of last resort' for some 26,000 people who could not escape the city.

Levée protection: The US Corps of Engineers had constructed the levées that surrounded the city of New Orleans and protected it from the river Mississippi and the Gulf of Mexico. Much of the city was up to 10 metres below sea level. The levées were built to survive a category 3 hurricane event but not a category 5. The US Congress had turned down plans to upgrade the defences to withstand a category 5 event and the budget for the US Corps of Engineers had been under pressure in the 10 years up to Katrina.

Building design and insurance: Any building in a potential hurricane zone must be built to tight regulations. Many of the buildings in New Orleans were built to withstand the hurricane-force winds, but this did nothing to limit the impact of flooding in the area.

> **Examiner tip**
> If asked to 'evaluate' the measures put in place to reduce loss of life or damage to property, remember to include both positive and negative issues. Your answer needs to be balanced.

Summary

- Hurricanes are storms with high winds (above $114\,kmh^{-1}$) and extremely low pressure (below 970 mb).

- Hurricanes always track from east to west and spin away from the equator. They usually happen in the late summer and autumn.

- There are six key factors necessary for hurricane formation.

- The structure of a hurricane is very different from any other weather system. Features include the eye, eye walls, rain bands and storm rotation.

- Hurricanes are huge systems spanning over 1,000 km and reaching up to 12 km into the troposphere.

- Hurricane Katrina was a category-5 hurricane that hit the coast of the USA on 29 August 2005, causing a huge amount of damage to the city of New Orleans.

- Katrina had a big impact on people and property in the southern USA — killing 1,836 people and destroying 300,000 houses, with damage costing an estimated $108 billion.

- Many of the protective measures put in place to reduce the loss of life and damage to property in the New Orleans area can be evaluated both positively and negatively.

Questions & Answers

In each AS Unit 1 Geography paper there are seven questions:

Unit structure

	Compulsory?	Marks (out of 90)	Exam timing (out of 90 minutes)
Section A			
Q1 Fieldwork skills You must take a summary report and a table of data into the exam with you. Questions will be asked in relation to how you continue to process the information from this table and on other fieldwork experiences.	Yes	30	30
Section B			
Q2 Rivers: short questions	Yes	12	12
Q3 Ecosystems: short questions	Yes	12	12
Q4 Atmosphere: short questions	Yes	12	12
Section C			
Q5 Rivers: essay question	Answer two from questions 5, 6 or 7	24 marks = 12 marks for each question	24 minutes = 12 minutes for each question
Q6 Ecosystems: essay question			
Q7 Atmosphere: question			

Examination skills

As with all A-level exams there is little room for error if you want to get the best grade. Gaining a grade A is not easy in AS geography so you need to ensure that every mark counts. The following table shows the minimum UMS (uniform mark scale) that you need to access particular grades.

Grade	AS Unit 1 (out of 100)	AS Unit 2 (out of 100)	Overall AS marks
A	80	80	160
B	70	70	140
C	60	60	120
D	50	50	100
E	40	40	80

Each of the two AS exam papers is 1½ hours. There are 90 marks available on each, which means that you get 1 mark per minute to work your way through the paper. The main reason why so many students struggle with this paper is that they fail to manage their time appropriately and as a consequence they do not have enough time left to answer the essays at the end in sufficient detail. If you find that you have time left over in this exam, the chances are that you have done something wrong.

Exam technique

Students often find it difficult to break an exam question down into its component parts. On CCEA exam papers, the questions are often long and difficult to understand, so you need to work out what the question is asking before you move forward.

Command words

To break down the question properly, get into the habit of reading the question at least *three* times. When you do this it is sometimes a good idea to put a circle round any command or key words that are being used in the question.

A common mistake is failing to understand the task being set by the question. There is a huge difference between an answer asking for a discussion and one asking for an evaluation.

The main command words used in the exam are as follows:
- **Compare** — what are the main differences and similarities?
- **Contrast** — what are the main differences?
- **Define** — state the meaning (definition) of the term.
- **Describe** — use details to show the shape/pattern of a resource. What does it look like? What are the highs, lows and averages?
- **Discuss** — describe and explain. Argue a particular point and perhaps put both sides of this argument (agree and disagree).
- **Explain** — give reasons why a pattern/feature exists, using geographical knowledge.
- **Evaluate** — look at the positive and negative points of a particular strategy or theory.
- **Identify** — choose or select.

Structure your answer carefully

Sometimes the longer questions on exam papers — for example, question parts for up to 6 marks or essay questions for up to 12 marks — can be an obstacle for students. Later in this section we will look at some questions and give more guidance on how you should structure your answers.

One simple approach to consider is drawing up a brief plan for your answer so that you know where it is going and how you will cover all of the main aspects of the question. For example, you could draw a box to illustrate each element needed within an answer and fill each one with facts and figures to support the answer, using the marking guidance to help you work out how much time to spend on each section.

Show your depth of knowledge of a particular place/case study

The essay questions on the exam paper are usually focused on giving you the opportunity to apply knowledge and understanding of case study material to a particular question. It is really important to show what you know here.

Examiners are looking for specific and appropriate details, facts and figures to support your case. The better you know and understand your case studies, the higher the marks you can potentially achieve.

About this section

A practice test paper with exemplar answers is provided. This will help you to understand how to construct your answers in order to achieve the highest possible marks.

In order to cover the range of potential questions in the 'fieldwork' section, two practice questions are included. Don't forget that you must take into the exam a summary of fieldwork and a table of data, which you attach to the exam paper. Question 1 will require you to answer questions on the particular fieldwork that you have carried out.

The essay questions in this section offer variable level answers or alternative questions.

Examiner comments

Some questions are followed by brief guidance on how to approach the question (shown by the icon ⓔ). Student responses are followed by examiner's comments. These are preceded by the icon ⓔ and indicate where credit is due. In the weaker answers, they also point out areas for improvement, specific problems, and common errors such as lack of clarity, weak or non-existent development, irrelevance, misinterpretation of the question and mistaken meanings of terms.

Question 1A Fieldwork skills

(a) Study the points below, which outline some important considerations made by a student when preparing for a geography fieldwork trip:
- **Transport to the site**
- **Accessibility of the site**
- **Safety equipment**
- **Suitable clothing for fieldwork**
- **Communication devices**

Select *one* of the planning considerations above and discuss its importance and role within your fieldwork.   (3 marks)

ⓔ 3 marks are awarded for an answer that deals with both the *importance* and the *role* of the selected factor and makes a convincing case, with appropriate reference to the individual fieldwork. 1–2 marks are given for a more simple response, which might fail to address either the importance or the role, with unconvincing reference to the individual fieldwork.

Student answer to question 1A

(a) When completing a fieldwork project in the sand dunes we had to think about specific safety equipment that would keep us safe on the site visit. We were worried about bad weather, so our teacher told us to make sure that we had a waterproof coat, waterproof trousers and a hat and gloves to keep us warm (it was a very cold, wet and windy day — so this advice was really important!). Also, we had to wear walking boots as the terrain was rough and this would give us more support. We brought waders and a safety throw rope as we were going to have to collect water from the sea to help with our experiment.

ⓔ **3/3 marks awarded**   This makes reference to three different aspects of safety equipment that were used for this particular fieldwork. It discusses both the role of the equipment and the importance of this within the context of their visit to the sand dune.

(b) Describe and explain *one* sampling method that was used within your fieldwork.   (6 marks)

ⓔ You can select from a range of different sampling methods, but most relate to *one* of either random, systematic, pragmatic or stratified sampling. 3 marks are for a general description of the sampling method and the other 3 marks are for an explanation of how the sampling method works in the context of this particular fieldwork.

(b) The aim of our fieldwork was to visit a total of 15 sites across the sand dune ecosystem at Magilligan Point. We were trying to see how the infiltration rate (how quickly 200 ml of water soaked into the soil) changes as you move away from the sea.

Our sampling method was to use systematic sampling to take measurements at sites along a line stretching back from the sea through the sand dune system. We took measurements at 15-m intervals along this line. We did this as other sampling techniques would not have been right for this coursework. We wanted to see how the infiltration changed through the sand dune and not just at random places. Pragmatic sampling might have worked too as the problem with systematic sampling was that sometimes the sites were not showing the changes that actually happened in the sand dune.

ⓔ **5/6 marks awarded** This is a good description of how the sampling technique was used within the context of this fieldwork. There is good depth about how systematic sampling was used to look at the changes across the sand dune system. The answer then goes on to explain why this sampling method was appropriate and even mentions how another sampling method might have been used. However, this explanation could have been more focused.

(c) (i) Choose *one* of the following statistical techniques that could be used to analyse some of your fieldwork data. Your chosen technique must fit in with the aim of your fieldwork.
- **Spearman's rank correlation**
- **Nearest neighbour analysis**
- **Mean, median, mode and range**

In the space provided, complete your chosen statistical analysis and show all your calculations clearly. If relevant, comment on the level of statistical significance. (Significance graphs and formulae are provided — see p. 12.)

(7 marks)

ⓔ The statistical technique depends on the chosen fieldwork, but it needs to be relevant to the aim/hypothesis of the investigation.

Spearman's rank correlation or nearest neighbour analysis:

- Accuracy of calculation (5 marks) (reduced to 3 marks if only seven ranked pairs noted)

- Statistical interpretation (2 marks)

- Maximum of 4 marks if selected technique is not appropriate for the testing of the aim/hypothesis

Mean, median, mode and range:

- Calculation of mean (2 marks)

- Calculation of median (2 marks)

- Identification of mode (1 mark)

- Calculation of range (2 marks)

(c) **(i)** Statistical technique selected: Spearman's rank correlation

Site (distance from the sea in m)	Rank	Time for infiltration/s	Rank	d	d^2
0	15	10.1	15	0	0
15	14	32.3	14	0	0
30	13	62.5	12	1	1
45	12	95.5	11	1	1
60	11	120	10	1	1
75	10	133	8	2	4
90	9	62	13	−4	16
105	8	156	7	1	1
120	7	132	9	−2	4
135	6	171.4	6	0	0
150	5	235	5	0	0
165	4	245	4	0	0
180	3	266	1	2	4
195	2	256	3	−1	1
210	1	262.5	2	−1	1
					$\Sigma d^2 = 34$

$$r_s = 1 - \left(\frac{6 \sum d^2}{n^3 - n} \right)$$

$$r_s = 1 - \left(\frac{6 \times 34}{15^3 - 15} \right)$$

$$r_s = 1 - \left(\frac{204}{3375 - 15} \right)$$

$$r_s = 1 - \left(\frac{204}{3360} \right)$$

$$r_s = 1 - 0.06$$

$$r_s = 0.94$$

This shows a strong positive correlation between the two variables. The result is significant at the 99.9% level. This is a very significant result.

ⓔ 7/7 marks awarded   This student has opted to complete a Spearman's rank correlation using the two data variables that are indicated in the aim of the fieldwork. The table is adapted for Spearman's rank data. Full working out of the formula is shown, with a final Spearman's rank result. The answer then comments on the relationship and on the significance of the result. There are no errors or inaccuracies here and the organisation of the task is logical, answering all aspects of the question.

(c) (ii) Explain your statistical outcome with reference to relevant geographical theory or concepts.

(6 marks)

ⓔ The answer must be based on *geographical* reasoning in relation to the calculated statistical result. The reasoning will depend on the fieldwork undertaken. There should be evidence of geographical theory or concepts to support the interpretation.

Level 3 (5–6 marks): Sound, relevant, geographical concepts are presented and discussed in an effective manner, using specialist terminology within written communication.

Level 2 (3–4 marks): An accurate discussion, but the answer may lack depth. There may be less evidence of geographical theory.

Level 1 (1–2 marks): Explanation is basic and lacks depth.

(c) (ii) Having completed the Spearman's rank correlation, I found that $r_s = 0.94$. This means that there is a strong positive correlation between the two variables. The result is significant to the 99.9% level, which means that the relationship did not occur by chance.

There is a geographical reason that connects the two variables. As you move away from the sea and move through the sand dune the soil structure in the sane dune begins to get more complex. This is because the vegetation on the sand dune gets more complicated — at the front of the sand dune the only plant that can survive is marram grass and this starts to hold the sand/soil together, but as you move further back into the sand dune ryegrass and moss start to develop and these help to add nutrients and humus into the soil and the soil can start to retain more moisture, making the infiltration rates slow down the further back that you go.

ⓔ **5/6 marks awarded** This is a good answer, which describes the relevance of the statistical technique in testing the relationship within the stated aim/hypothesis. There is also good reference to geographical concepts and theory to back up the technique used. Specialist terms are used and specific knowledge in relation to seral succession is shown.

(d) Describe and evaluate *one* way in which you could extend your fieldwork study to explore your aim further and improve your geographical knowledge.

(4 marks)

ⓔ Answers will vary depending on the fieldwork chosen. 2 marks are awarded for description of a valid possible extension to the fieldwork. The other 2 marks are for an evaluation of how this extension might improve the fieldwork experience.

(d) The main aim of our fieldwork was to look at the relationship between distance from the sea and the infiltration rate. We realised when we had completed the 15 sites that we had really only gone about halfway through the sand dune. We had not made our way through all of the different changes in the sand dune and had not reached the climax vegetation levels. Therefore, to improve our fieldwork we should have continued until we had reached the back of the sand dune — though this might have meant another 20 site measurements, which would have taken a long time. However, we would have had a lot more data with which to make sure that our hypothesis was accurate.

ⓔ **4/4 marks awarded** This makes clear reference to the aims stated in the report and then describes how the fieldwork could be improved further by visiting additional sites. The answer is developed further with some evaluation of how this might be positive (improving accuracy) but also negative (a lot more sites needed = more time).

(e) Describe *one* way in which your fieldwork location proved to be suitable for carrying out the stated aim in your fieldwork report. (4 marks)

ⓔ You are expected to reflect on the location of your fieldwork (as noted in the report) and should attempt to justify the suitability of the place in the light of the stated aim. You should note the aim and link this to a discussion about how suitable the place was. 3 or 4 marks are awarded for a valid, well-argued answer that demonstrates good knowledge of the chosen fieldwork site and links this with the stated aim/hypothesis. A more limited answer, for 1 or 2 marks, might be down to failure to discuss the suitability of the site or link it to the aim/hypothesis.

(e) In our study we were aiming to see if there was a link between the infiltration rate and the distance that you move away from the sea into a sand dune. In class we had learnt about the different changes that a sand dune goes through as it develops and so we looked for a sand dune system that would allow us to see some of these changes. Magilligan Point was a really good dune to study as we were able to see pictures of what the dune used to look like 200 years ago and we could see how this was different from today. The sand dune system was accessible and was a safe environment for us to record our results.

ⓔ **3/4 marks awarded** There is some good depth to this answer. It refers well to the aim of the study and explains why this particular location might have been suitable, but to get the final mark it should have made the link between the aim and the location more detailed. For instance, why was this a better location for taking infiltration measurements than another place? Conversely the answer could have argued why using a different location would have given better or more accurate results.

Question 1B **Fieldwork skills**

(a) With reference to sampling and risk assessment, discuss how these might be considered by a geographer when planning a field study.

(6 marks)

ⓔ 1 mark is awarded for a general understanding of the task, 1 mark for a clear and specific link to fieldwork, and 1 mark for an explanation of how the task was completed. Each task is worth 3 marks.

Student answer to question 1B

(a) Sampling: In our fieldwork study at Magilligan Point, Co. Derry, we used a systematic sampling technique. To do this we identified 15 different sites that we would sample. These sites went from before the foredune through to the grey dune. Each site was 15 metres further from the sea than the last site.

Risk assessment: For our fieldwork we identified the key risks at Magilligan Point — for example, the lack of shelter and strong winds from the sea. The local weather forecast was then checked before our research study so that we could bring suitable equipment and clothing to ensure safety — for example, waterproof coats and gloves. Risk assessment was also carried out by identifying key hazards on the ground and making sure that people carried the equipment safely.

ⓔ **6/6 marks awarded** Both tasks are dealt with in good depth. A general understanding of both tasks is shown and a clear link is developed specific to the fieldwork outlined in the fieldwork report. There is some discussion about how the sampling and the risk assessment were carried out in this particular case. This answer is worth the maximum 3 marks for each task.

(b) Describe *one* of the primary data collection techniques that were used in your fieldwork to produce the results in your submitted table.

(3 marks)

ⓔ You must select and write about a primary source of data collection that is referenced in your submitted table. 3 marks are awarded for a detailed description of the data collection technique with explicit and detailed reference to the equipment used (if relevant). 1 or 2 marks are awarded if the description of the techniques lacks depth and reference to fieldwork.

(b) To measure the time taken for 200 ml of water to infiltrate the soil we needed an infiltrometer, 200 ml of water, a measuring cylinder and a stopwatch. The vegetation was cleared so that we were measuring the time taken for the water to infiltrate the soil and not the vegetation. The infiltrometer was placed firmly in the soil so that water did not escape through the sides. 200 ml of water was then poured in and the time taken for it to infiltrate was recorded. The data were collected accurately as we did it three times and took an average.

ⓔ **3/3 marks awarded** This is a good description of the technique used in the fieldwork. It makes explicit reference to the equipment used to generate the fieldwork table and shows how the equipment was used. This answer scores full marks.

(c) (i) Using some, or all, of the information from your fieldwork table and report, draw a graph relevant to the aim of your fieldwork.

(7 marks)

🄴 Your graph must be accurate and relevant to the aim of the fieldwork, and you must use data from your fieldwork table. 1 mark is available for the title: you must state clearly each of the variables presented. 2 marks are for conventions: axes must be labelled (variables and units of measurement), a key must be included and scaling (if appropriate). 3 marks are available for accuracy: precision of values, and 1 mark is available for the method: selection of appropriate graph for the task. (Graph paper will be provided in the exam.)

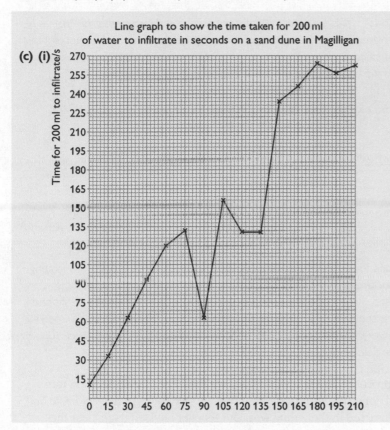

Line graph to show the time taken for 200 ml of water to infiltrate in seconds on a sand dune in Magilligan

(c) (i)

🄴 **5/7 marks awarded**   The title is long but it generally describes the content of the graph. Conventions have been followed but one of the axes is not labelled fully so only 1 mark can be scored. Some of the values are not as precise as they could have been and this means only 2 marks can be awarded for accuracy. The method selection (a line graph) is acceptable. Take care to draw graphs quickly and carefully.

(c) (ii) Describe and explain *one* geographical factor that might have influenced the results displayed on your graph.

(4 marks)

🄴 Again, the answer will depend on the particular fieldwork studied. The factor selected must be geographical and can relate to either human or physical factors. 4 marks are available for a thorough geographical discussion (interpretation of the graph) that clearly refers to and explains the results generated in the graph. 1 or 2 marks will be awarded for a less-well-argued discussion.

(c) (ii) The amount of vegetation. As you move back through the psammosere, the amount of vegetation became increasingly thick and dense. It became very difficult to remove the vegetation to measure the infiltration rate. At some of the sites it may have been impossible to fully clear the site of vegetation and this could have increased the time taken to infiltrate. The more complex the vegetation through the dune, the more humus and leaf litter will also make the soil more complex.

ⓔ **3/4 marks awarded**   This is a decent attempt to describe and explain the role that vegetation might have played in the development of the ecosystem and how the vegetation has made an impact on the infiltration rate. The answer goes into some depth but could make more reference to how vegetation had affected the actual infiltration rates at each site.

(d) Select two of the factors shown below and explain how they might have influenced your results and final conclusions.
- **Time of year**
- **Fieldwork equipment**
- **Weather conditions**
- **Time of day**
- **Group organisation**
- **Human influence**

(6 marks)

ⓔ You need to evaluate the influence that each of two factors might have played in generating the results and final conclusions to the fieldwork. 3 marks are available for one factor if the argument is coherent and shows a good understanding of how the factor influences results and conclusions. The answer should clearly reference the fieldwork. 1 or 2 marks will be awarded for a less commanding answer that might not explain the influence of both results and conclusions, or might not reference the fieldwork.

(d) Fieldwork equipment: in our fieldwork study we measured the infiltration rate across the sand dune. The infiltration ring that we were using was very basic and sometimes it was used in the wrong way. Our teacher advised us to clear the vegetation and then try to sink the ring into the sand/soil as much as possible but this was not always easy and sometimes water leaked out of the side, and this might have affected the results and the conclusions. It would have been better to use a steel, double-ringed infiltrometer as this might have given more accurate results.

Human influence: sand dunes show lots of human influence — there are paths, signs, litter and even an old castle in the dunes. There is no doubt that humans have played their part in the history of the dunes, which means that the dunes that we see and the infiltration rate might be influenced by people.

ⓔ **4/6 marks awarded**   The explanation about the fieldwork equipment is much stronger than that about human influence. The discussion of fieldwork equipment makes reference to both the influence on results and on conclusions, and scores the full 3 marks. The second factor is less well explained and needs further development to show how human interference might have affected both the results and the final conclusions.

> **(e) Discuss the purpose of statistical analysis as part of any fieldwork study and explain briefly why your chosen statistical method was selected as suitable for your fieldwork.**   (4 marks)

ⓔ Statistical analysis helps students to test their hypotheses in any investigation. It helps geographers to reach valid, reliable geographical conclusions. It allows a lot of data to be compared in a meaningful and concise manner. For this question, 2 marks are available for an awareness of the purpose of statistical analysis in any fieldwork investigation and 2 marks are for a justification of the chosen method. You need to make reference to aim/hypothesis and show why this technique was suitable.

(e) Statistical analysis in a fieldwork investigation provides a method of processing, calculating and interpreting raw data into a result, which can help to prove or disprove the hypothesis. Spearman's rank was chosen to give a result, in this case, $r_s = 0.94$, which allows us to interpret the degree of correlation, which was highly significant. This helps prove our hypothesis, in relation to infiltration rates, which stated 'The time taken for water to infiltrate will increase back through the psammosere'. Our Spearman's rank showed that this was a suitable measure as the relationship between the two variables could be measured and the strength of relationship could be commented on.

ⓔ **4/4 marks awarded**   The first element of the question is answered well and some of the reasons why statistical analysis can be important are explained. The reasons why Spearman's rank is appropriate in this case study are explained and the outcome of using this technique is shown clearly and linked to the aim of the study. This is a well-argued answer that scores full marks.

Question 2 **Rivers short questions**

(a) With the aid of a well-annotated diagram, describe and explain the role of deposition in the formation of a floodplain.

(6 marks)

ⓔ Level 3 (5–6 marks): The answer uses a well-annotated diagram and specialist terminology to explain the formation of the floodplain and associated river processes.

Level 2 (3–4 marks): The answer might use a less-well-annotated diagram of the feature and/or a less detailed explanation of its formation. Alternatively, an accurate and detailed explanation with no diagram might be marked at this level.

Student answer to question 2

(a) There are two main sources of deposition. The first are called point bar deposits, which are deposited on the lower-energy inside bends of migrating meanders. As the meanders weave across the floodplain, they leave these point bar deposits all across the floodplain. Secondly, if a river floods onto the floodplain, it experiences a lot of friction as it comes into contact with the ground and the vegetation on it. This friction slows the velocity of the river, causing deposition to occur. The largest particles need most energy to transport them, so they are deposited first (forming levées along the edge of the channel), while the smallest particles are carried furthest across the floodplain.

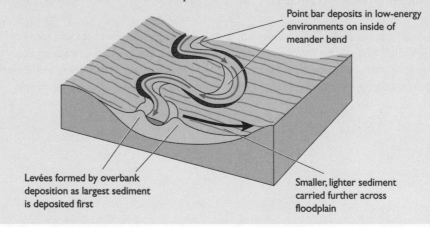

Point bar deposits in low-energy environments on inside of meander bend

Levées formed by overbank deposition as largest sediment is deposited first

Smaller, lighter sediment carried further across floodplain

ⓔ **6/6 marks awarded** This detailed answer shows good understanding of the depositional processes at work in the formation of a floodplain. There is good use of terminology and a well constructed and annotated diagram is included.

(b) Study Resource 1, which shows the Hjulström curve. After a rainstorm, the discharge in a river falls from 100 cm s⁻¹ to 1 cm s⁻¹. Use the Hjulström curve to describe and explain how this impacts on the river's load.   (6 marks)

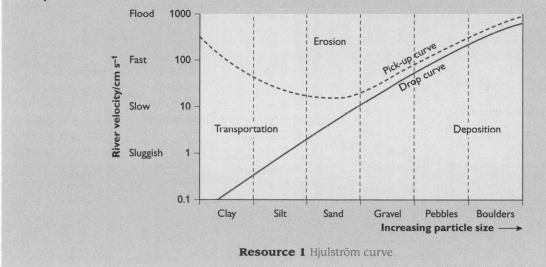

Resource 1 Hjulström curve

e Level 3 (5–6 marks): A detailed and thorough answer, which makes effective use of the resource to discuss the changes over time.

Level 2 (3–4 marks): A general but accurate answer, which discusses how the load changes over time. Use of the resource may be restricted.

Level 1 (1–2 marks): A limited answer, which fails to address the changes over time and/or makes no meaningful reference to the resource.

(b) At the start of the time period, when the velocity is at 100 cm s⁻¹, the river has enough energy to transport a large range of sizes of load, from clay up to gravel-sized particles. However, during this time period, as velocity drops from 100 cm s⁻¹ to 1 cm s⁻¹, the river's energy drops and so its carrying capacity also drops. This means that the river begins to deposit more and more load, starting with the largest particles — gravel. The main reason for this is that gravel is larger and so requires more energy to transport than smaller particles such as sand. As velocity drops further, however, the smaller particles begin to be deposited: sand is deposited at around 10 cm s⁻¹ and silt at around 1 cm s⁻¹. Below 1 cm s⁻¹, the smallest clay particles continue to be transported as they are so light.

e **6/6 marks awarded**   This detailed and thorough answer clearly addresses the changes over time. Resource use is detailed, with plenty of figures quoted, and the changes are clearly explained.

e **The focus of this question is on how the transportational _and_ depositional processes change during the time period. As discharge and therefore river energy fall, the load begins to be deposited, starting with the largest particles.**

Question 3 **Ecosystems short questions**

Study Resource 2, which shows the change in vegetation primary succession from bare rock.

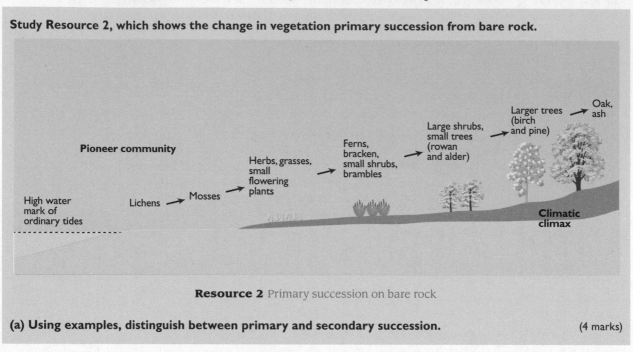

Pioneer community

High water mark of ordinary tides

Lichens → Mosses → Herbs, grasses, small flowering plants → Ferns, bracken, small shrubs, brambles → Large shrubs, small trees (rowan and alder) → Larger trees (birch and pine) → Oak, ash

Climatic climax

Resource 2 Primary succession on bare rock

(a) Using examples, distinguish between primary and secondary succession. (4 marks)

ⓔ 2 marks are for demonstrating clear understanding of the contrasting definitions and 2 (×1) marks are for an example of each.

Student answer to question 3

(a) Primary succession happens when plants colonise and grow on a surface which did not have vegetation on it before such as bare rock on a newly formed volcanic island. In contrast, secondary succession occurs on surfaces where the vegetation has been destroyed, such as when fire creates a clearing in a forest.

ⓔ **3/4 marks awarded** This communicates the clear distinction between the two processes and quotes good examples. This is a case of the student reading the question carefully. See the few extra seconds it takes to read the question carefully as a vital investment and not a waste of your time.

(b) Explain *one* soil change and *one* microclimate change that occurs to produce the succession shown in Resource 2. (4 marks)

ⓔ As succession occurs, the abiotic environment is progressively changed. You should explain one change to the *soil* (e.g. depth, humus, moisture, pH) and one to the *microclimate* (e.g. wind speed, temperature).

(b) The soil becomes deeper due to the addition of more organic matter from the biomass and increased weathering of the bedrock to provide mineral content. Also, the microclimate improves, with the vegetation causing frictional drag on the wind and reducing surface wind speeds.

ⓔ **4/4 marks awarded**   This is clearly focused on the requirements of the question and shows a clear understanding of the two changes selected.

(c) **With reference to a named small-scale ecosystem you have studied, discuss the role of decomposers in the ecosystem.**   (4 marks)

ⓔ 3–4 marks are awarded for an answer that identifies decomposers and explains their role. An ecosystem must be named to achieve these marks. An answer that does not fully explain their role or one that does not make reference to an actual ecosystem would be awarded 1 or 2 marks.

(c) In the deciduous forest in Breen Wood there are various decomposers such as earthworms, bacteria and fungi. They operate at each trophic level to break down and cause the decay of the organic matter not consumed by other organisms, including excrement and bones. This means that they play an important part in the recycling of the organic nutrients within Breen Wood.

ⓔ **4/4 marks awarded**   This full answer provides a spatial context, identifies named decomposers and explains their role.

Question 4 **Atmosphere short questions**

(a) Study Resource 3, which shows some of the responses to Hurricane Floyd in the USA in September 1999. Use the resource and your own case study material to discuss how hurricane protective measures can be used to reduce the loss of life and damage to properties. (6 marks)

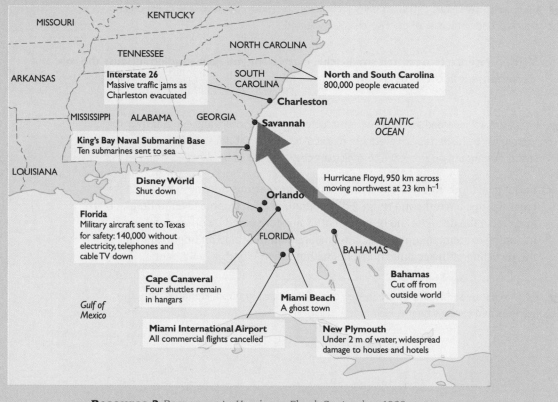

Resource 3 Responses to Hurricane Floyd, September 1999

ⓔ A good answer will make reference to some of the features on the resource *and* some case study material that looks at the impact on both people and on property.

Level 3 (5–6 marks): The answer discusses a range of protection measures/responses, making reference to the resource *and* your own hurricane case study material. Discussion should involve how the measures have been able to reduce the loss of life and/or damage to property.

Level 2 (3–4 marks): There is less detail about the measures taken to reduce the death toll and/or damage to property. The case study or resource might be left out at this level.

Level 1 (1–2 marks): A basic answer, which lacks explicit reference to a case study and/or the resource material.

Student answer to question 4

(a) When Hurricane Floyd hit the USA in 1999, the authorities were able to instigate an evacuation programme to get people living near the coast and in vulnerable areas to move to places of safety. Over 800,000 people were moved from North and South Carolina, flights were cancelled and Disney World was closed to keep people safe.

The same things happened before Hurricane Katrina hit — on 28 August 2005 the Mayor of New Orleans announced the first evacuation of the city. However, one problem with this evacuation was that many of the people in the city were very poor and had no means of transport to get out of the city.

When Hurricane Floyd hit, some of the naval submarines had been sent out to sea to minimise damage and other aircraft had been flown to Texas and the space shuttles were kept in their hangars. In New Orleans, people were able to board up their properties as they were given a warning from the Early Warning System — though people did not really expect the flooding that took place.

ℯ **5/6 marks awarded** This makes some use of the resource to answer both aspects of the question: how hurricane protection measures can reduce the loss of life and damage to properties. It also mentions the student's case study material — the impact of Hurricane Katrina on New Orleans — but could have gone into a little more specific detail.

(b) Outline one factor that might influence the global temperature pattern. (3 marks)

ℯ This requires you to choose and outline one control or influence on the global temperature pattern, for example ocean currents, continentality, altitude or prevailing winds.

3 marks are awarded for an answer that develops a detailed description of *one* influence. There should be reference to the differences in global temperature, with places near the poles being much colder than at the equator. 2 marks are awarded for an answer that goes some way to describe the influence of one factor but which maybe does not explain the global variation in enough depth. An answer that does identify a valid factor but is vague and does not develop specific detail is worth just 1 mark.

(b) Ocean currents can have a big impact on the global temperature pattern. The sea can absorb heat over a long period of time compared with the land and it retains its heat for a much longer time, which means that the ocean currents move blocks of warm water and air towards colder areas. This movement around the globe moves warm water from the equator towards the poles. A good example is the North Atlantic current, which moves warm water from the Caribbean to Western Europe. This keeps the UK much warmer in the winter than inland Europe.

ℯ **3/3 marks awarded** This identifies a valid factor, explaining it in good depth and making reference to the fact that some areas of the Earth are warmer than others due to this factor.

(c) Describe the difference between horizontal and vertical heat transfer when talking about the global energy balance. (3 marks)

ⓔ 3 marks are awarded for sound awareness of the distinction between both transfer methods and valid reference to global energy balance. No more than 2 marks are awarded if examples are not given when talking about both transfers or if no reference is made to global energy balance.

(c) Horizontal heat transfer transfers heat towards or away from the equator. An example of this is ocean currents or the wind. Vertical heat transfer transfers heat from ground level into the upper atmosphere or troposphere, for example radiation or convection currents.

ⓔ **2/3 marks awarded** This gives good definitions of the two different concepts but needs further development by giving more detail about how these might influence the global energy balance.

Question 5 Rivers essay

With reference to a large-scale drainage basin or delta you have studied, discuss the physical and human causes of flooding.   (12 marks)

ⓔ You should name a large-scale drainage basin or delta for Level 3 and discuss a range of both human and physical causes for the flood event or events named. Physical causes might include seasonal climatic change, for example snow melt or higher-than-average rainfall. Human factors could include deforestation, urbanisation, dam building or farming practices.

Level 3 (9–12 marks): The answer shows a good balance between physical and human factors, describing and explaining a range of causes of flooding. The case study is named and there are good references to case study details throughout. Sound understanding is communicated effectively using specialist terminology.

Level 2 (5–8 marks): The answer describes a range of human and physical factors, but with limited reference to case study detail. Alternatively, the answer is unbalanced, focusing too much on either physical or human. The answer may lack depth and fewer specialist terms are used.

Level 1 (1–4 marks): The answer may cover only one of the two types of cause identified in the question. The answer is superficial with no meaningful case study reference. The level of written communication may also be poor.

Student A answer to question 5

The Mississippi 2011 floods were caused by a variety of physical and human factors.

The first factors I will look at are physical or climatic factors. The first of these is heavy snowfall in the upper Mississippi and Missouri tributaries to the northwest where snowfall in the 2010/11 winter created a snow pack 60 cm deeper than average. This then melted in the spring and a large volume of water flowed into the rivers, creating a flood peak discharge that reached 15 m at Thebes, which was the largest since records began. At the same time, in the northeast of the drainage basin around the Ohio River tributary, a second physical factor of record-breaking rainfall was in operation. During April, a series of four intense rainstorms produced six times the average monthly rainfall in this area. This also resulted in a flood peak travelling down the river towards Cairo where the confluence between the Ohio River and the Missouri/upper Mississippi Rivers is located. These two flood peaks met at this confluence, creating a massive flood peak heading into the downstream area of the lower Mississippi basin.

Once the physical causes had triggered the flood, a variety of human factors increased the scale of the flooding. The first of these was related to human management of the river. The 3,000 km of levées that were built around many of the cities on the Mississippi River since the 1927 flood as well as protecting some areas from flooding actually can make flooding worse elsewhere. For example, the increased discharge behind the levées can increase the river's erosional power and can cause the levées to fail. So, to prevent this from happening to the city of Cairo, the levée at Birds Point was deliberately blown up to get rid of some of the water from the river. This breach flooded 13,000 acres of farmland and 100 homes.

A second human cause was urbanisation. Following the 1993 flood, FEMA bought back 12,000 at-risk homes on the floodplain to create a safe flood zone. However, since then, pressure for urban development led to people building on the vulnerable floodplain. For example, there is now more urban development on the floodplain around St Louis than there was before 1993. This not only puts those homes at risk, but it increases demand for more levées, which in turn can fail and cause worse flooding elsewhere.

(e) **11/12 marks awarded**   This answer includes a range of physical and human factors (two of each). The case study is named and there is good spatial detail throughout, although one or two more facts would have been preferable. There is good use of terminology (such as flood peak and confluence) and the quality of written communication is very effective in that the answer clearly demonstrates a depth of understanding of the issues.

Student B answer to question 5

Mississippi Flood 2011

The physical causes were firstly the winter snow pack in the upper west course of the Mississippi River, which caused the record snow melt to flow down the Mississippi River and meet the flood waters coming from the Ohio River tributary. Another physical cause was four intense rainstorms that occurred in April 2011 and brought intense rain and flooding. The last of the four rainstorms occurred between 25 and 28 April 2011. This meant the soil was saturated and could not allow any more infiltration, resulting in intense flooding.

There were also human causes. Firstly urbanisation. As many people began to move to the Mississippi basin, large cities began to grow, such as St Louis which is the sixteenth largest city in the USA and the fourth largest in the mid-West. This urbanisation meant lots of impermeable surfaces such as concrete. The storm drains used in St Louis also meant lag time to the river was reduced, increasing its peak discharge. Lastly, the drainage of 26 million acres of farmland meant that an area of natural storage was removed. The US Army Corps of Engineers also blew up the Birds Point levees in Wyatt, meaning the river's capacity was reduced and the Morganza spillway was opened, flooding many areas in Louisiana County. By 15 April, nine flood gates had been opened.

(e) **8/12 marks awarded**   Although many valid factors are proposed, the answer could include a deeper level of explanation of factors, more specialist terminology in places and some specific case study depth/details, for example rainfall figures for the Ohio River area. This greater depth could be achieved by tackling a slightly smaller range of factors but dealing with them in more detail.

Question 6 Ecosystems essay

> **Describe and evaluate the soil conservation methods used in the management of a mid-latitude grassland ecosystem you have studied.**   (12 marks)

ⓔ Both key elements need to be addressed: knowledge and understanding of the management strategies, along with an awareness of their effectiveness/limitations. Appropriate case study detail should be included. Although not required, you may incorporate diagrams to help describe the operation of the strategies.

Level 3 (9–12 marks): A detailed answer, which addresses both aspects of the question and introduces a wide range of soil conservation methods. Case study detail and specialist terminology are included.

Level 2 (5–8 marks): Strategies are mentioned but detail is lacking in either description or evaluation. Reference to spatial detail is more limited. A narrower range of techniques may be discussed.

Level 1 (1–4 marks): The answer shows limited knowledge of the strategies, and inaccuracies may be evident. Reference to case study detail may be largely ineffective or missing entirely.

Student A answer to question 6

In the mid-latitude grasslands of the North American prairies, there have been various soil conservation methods introduced, particularly to help reduce the level of erosion.

Firstly, mulching has been introduced. This involves residue-conserving practices, which means leaving behind the stalks of the harvested crop. It is estimated that 1,500kg per hectare of residue is needed to prevent the major wind and water erosion that occurs.

Secondly, green manure crops have been introduced. This involves planting green crops such as peas and lentils to increase the nitrogen levels of the soil. This is an alternative solution to the fallowing of land, which has been extremely damaging in the prairies. It prevents the land being left bare and therefore vulnerable to erosion.

Shelter belts have been introduced to conserve the soil in the mid-latitudes grasslands. These are densely populated hedges of trees, which prevent erosion of the soil.

Another soil conservation method that has been introduced is contour ploughing. This involves ploughing along contour lines to avoid gullies, which increase fluvial erosion. This has been introduced in the less flat areas of the prairies such as Nebraska.

Strip farming is another soil conservation method that has been introduced. Crops are grown in strips that alternate with strips of fallow, reducing the impact of soil erosion. This has been used in Saskatchewan, where there's a northwest wind direction, so the crops are grown north and south to reduce aeolian and fluvial erosion.

Finally, there has been a programme called the 'Native Prairies Protection Program' introduced, which rewards landowners for protecting and managing the mid-latitude grassland in a way to reduce erosion brought about by poor farming practices.

ⓔ **7/12 marks awarded** This answer does give a wide range of techniques and includes some good spatial detail, but it makes little attempt at evaluating the effectiveness of the management, as the question clearly requires. Read the command phrases carefully in the questions and be clear on what each one wants you to do.

North American prairies

Management strategies used are mulching. This is to maintain crop residue by leaving stalks behind from crop plants to protect the soil from wind erosion after the plants are harvested. Also shelter belts, which is growing trees on edges of fields at 90 degrees to the prevailing winds, to reduce the distance wind has to move across the fields and acting as wind breaks. Farmers are paid $40 per acre per year for 10 years or $70 per acre per year for 21 years.

Green manuring is an alternative to fallowing. It involves growing crops not for harvest but to work into the ground by tillage. These plants return nutrients to the soil. Prairies use legumes because they fix nitrogen from the air into the soil. Strip farming is growing crops in strips alternating between bands of fallow. This acts as a wind break and reduces soil erosion by the wind. Contour ploughing is the final management strategy. It means ploughing along contour lines to avoid making gullies that will increase water erosion and so reducing nutrient loss.

These strategies have been successful to a certain extent. Soil erosion has been reduced — from 1982 to 2007 it fell by 43%. In the southern plains region, erosion has gone down from 12.5 tons per acre per year in 1982 to 8.8 tons per acre per year in 2007. However, erosion is still occurring. 99 million acres were eroded unsustainably in 2007 with a loss of 827 tons of soil. In light of this, many have called for a more sustainable management strategy like returning the prairies to their natural condition. An example of this is Buffalo Commons, where 360,000 km^2 was to be returned to natural grassland. Initially, this idea was controversial but it has grown in popularity.

e **6/12 marks awarded** This answer attempts the two main elements of description and evaluation. The evaluation is more detailed with good spatial context. But the strategies are more variable, with some lacking detail of description and lacking sufficient spatial context.

Question 7A Atmosphere essay

Discuss the formation of a mid-latitude frontal depression and then make reference to your case study to analyse the impact of the depression on people.

(12 marks)

ⓔ There are two aspects to this question that need to be addressed: you need to show an understanding of the formation processes of a depression and then demonstrate some of the effects that a depression like this might have on people. Formation processes could include reference to air masses involved in the formation, or the development at the polar front. Impacts on people might include reference to heavy rainfall, strong winds, poor visibility and thunderstorms.

Level 3 (9–12 marks): The answer shows detailed understanding of the formation processes. Relevant case study material is evident and can be used to show the impact that a depression might have on people.

Level 2 (5–8 marks): There is a general understanding of the formation of a depression and some explanation of impacts. However, case study material might be limited.

Level 1 (1–4 marks): A basic answer with little, if any, case study material.

Student answer to question 7A

Depressions are one of the main weather systems to affect the British Isles. They are formed out in the North Atlantic Ocean when cold polar maritime air from the north moves south and joins up with some warm tropical maritime air moving north. When the air meets, this creates a front and frontal rainfall is caused. The lighter warm air is forced to rise over the slower-moving, more dense cold air. The depression continues to move from the west towards the east and brings low pressure, cloud, rain and windy conditions to the UK.

Depressions can have quite a big impact on people. One depression that impacted the UK was the great storm on 2 February 1994. The pressure fell to 954 mb and this caused storms across the country. Gale-force winds up to 100 mph were felt across the UK. In Northern Ireland, the NIE recorded the biggest loss of power on record as 400,000 homes were left without any electricity supply due to the strong winds. Travel links were disrupted — ferries were cancelled, airports were closed and trains and lorries were disrupted by the winds and flooding due to the heavy amounts of rainfall.

ⓔ **8/12 marks awarded** This student has presented some good information and development of the facts of the particular case study, while addressing the two elements of the question. There is good, detailed discussion of the formation processes and also some development of the different impacts that a depression has had in relation to a particular case study. However, there could have been a little more specific information added here to take the answer into Level 3.

Question 7B Atmosphere essay

With reference to your case study of a hurricane, evaluate the protective measures used to reduce the loss of life and damage to property.

(12 marks)

ⓔ You must name a specific hurricane event and then evaluate the protective measures that were taken to reduce the loss of life (impact on people) and to reduce the impact of damage to property. Evaluating demands more than a simple discussion — there should be reference to strengths and weaknesses of measures and some critical reflection.

Level 3 (9–12 marks): The answer shows detailed understanding of the hurricane and the range of measures is outlined and evaluated. There is good depth and explanation.

Level 2 (5–8 marks): There is a general understanding of the case study. The level of knowledge of the case study might be limited. There might be some attempt to evaluate protective strategies.

Level 1 (1–4 marks): A basic answer with little, if any, case study material.

Student answer to question 7B

On 29 August 2005 Hurricane Katrina hit the city of New Orleans on the southern coast of the USA. The hurricane brought high winds and rainfall and created a storm surge, which burst the banks of the levées that protected the city from flooding. After the floodwaters had gone down, many of the people who lived in the area were critical of the protective measures that had been taken to reduce the loss of life and damage to property.

Measures taken to reduce the loss of life

Positive: In the USA there are organisations like the National Hurricane Centre that will track hurricanes and work out where they are going to hit landfall. This means that only people who really need to be evacuated will be moved. New Orleans did have its own disaster plan, which had been practised many times. The emergency services had planned for a disaster like Katrina so they should have been ready for whatever happened.

Negative: FEMA — the national USA organisation for managing disasters —reacted slowly to the disaster. It did not take decisions quickly enough in the early stages of the emergency. Even though the plans had been rehearsed, many of the poor people in the centre of the city were left to themselves as they had no way of getting out of the city. 25% of the population had no car, plus the final decision to evacuate the city was only made 19 hours before the hurricane arrived. This meant that there was not enough time to get everyone out of the city.

Measures taken to reduce damage to property

Positive: Most of New Orleans is below sea level. Levées, built by the US army, were designed to protect the city in the event of a hurricane or a flood. However, these were not built strong or high enough for the impact of Katrina — a category 5 hurricane. Though many people note that things would have been a lot worse if these measures were not taken at all.

Negative: The levées were ruptured very quickly and this caused a mass of floodwater to move through the city. Buildings were not designed to withstand both the high winds and the high water levels. The government was more concerned with saving lives; they did not care that property was being damaged.

ⓔ **9/12 marks awarded** There is some good development and discussion about the protective measures used, with reference to Hurricane Katrina and New Orleans. There is a nice balance between the positive and negative and between property and people. However, more detail could have been given in relation to property (especially the negative impacts). It would have been good to see more facts and figures throughout the discussion as well.

Knowledge check answers

1 Because primary data are recorded by the person doing the fieldwork, it is easier to assess how accurate and useful the data and results will be. If fieldwork is planned carefully, the primary data can make a big difference in the interpretation of results. However, collecting primary data can be time consuming and expensive (especially if specialist equipment is required) and can sometimes not be as accurate as it might be.

2 Infiltration refers to water entering the ground surface and transferring into the soil store, whereas percolation refers to the movement of water down through the soil profile until it reaches the groundwater store.

3 During the summer trees have leaves, which increase the store of interception. This delays the water reaching the ground by the transfer of stemflow and so the ground is less likely to have its infiltration capacity exceeded. This results in less overland flow. In contrast, during the winter there is less interception and so the water arrives at the ground more quickly and more overland flow results.

4 As a drainage basin experiences increased urbanisation, lag time is reduced and peak discharge increases. This is due to the increase in impermeable surfaces and the introduction of drains and sewers. These both increase overland flow and provide an efficient route for the water to the river, and so the water arrives at the river more quickly.

5 Corrasion/abrasion, hydraulic action and attrition are affected by discharge and energy levels as they are physical processes. If a river has more energy, these processes are more efficient. In contrast, corrosion is a chemical reaction between weak acids in the river water and the rock over which the river flows.

6 In cross section, meanders are asymmetrical. On the outside of the bend they have a steeper bank (which is often undercut by erosion) known as a river cliff. On the inside of the bend, they have a much more gently sloping bank of point bar deposits, known as a slip-off slope.

7 The rate can be affected by how steep the sea floor gradient is. The steeper the gradient, the slower the rate of formation as the sediment being deposited is more likely to fall down the slope under the influence of gravity.

8 Trophic level 2 consists of the herbivores. These animals feed on the plants and so are called heterotrophs or primary consumers. At trophic level 3 we find the carnivores that feed on the herbivores from trophic level 2, so they are heterotrophs or secondary consumers.

9 Energy flows, i.e. it *enters* via photosynthesis, *moves through* the trophic levels and is finally *outputted* as heat via respiration. In contrast, nutrients cycle, i.e. they move *between* the stores (biomass, litter and soil) *within* the ecosystem.

10 The three aspects that change are vegetation, soils and microclimate.

11 In the psammosere at Portstewart Strand, the soil quality improves considerably as the pH changes from being very alkaline in the embryo dunes to around 6.5 in the grey dunes. Species diversity increases from mostly sand couch in the embryo dunes to a range of stratified plants in the grey dunes, including bedstraw and bird's-foot trefoil.

12 The effects of leaching in a mollisol are limited and only a small proportion of the nutrients are removed from the soil by this means. This is because precipitation levels only slightly exceed evapotranspiration levels. In fact, soil moisture levels are so low in the summer that water moves upwards into the profile via capillary action. This can lead to nodules of calcium carbonate in the lower Cca horizon of the soil.

13 Monoculture has had a massive impact in the prairies of North America. The soils have experienced significant soil erosion. Ploughing of the soil breaks up the surface mull humus, damaging the soil's structure. This makes it more at risk from erosion. Following harvesting, the soil is often left bare, leaving it vulnerable to wind and water erosion. One of the most significant examples of this is the Dust Bowl in the 1930s where winds removed up to 75% of topsoil. Despite this, portions of land were still left bare after the Dust Bowl. Around 9.5 million hectares of land were left bare in 1981, an increase of 42% since 1931.

Vegetation is also affected by monoculture. For example, there has been a significant drop in biodiversity, with the range of crops like prairie dropseed and prairie dock being replaced by a single cereal crop. The US government in 1990 estimated that over 350 plant species were under threat in the prairies.

14 The three processes that can transfer heat vertically in the atmosphere are radiation (where the land radiates heat back out to space through long-wave radiation), convection (where warm air is forced to rise as part of convection currents, and rising warm air is replaced by colder, descending air) and conduction (where the energy is transferred through contact).

15 Continentality influences the pattern of temperature because areas that are further inland experience a much bigger range of temperatures than those on the coast. It is generally warmer in the summer and colder in the winter inland.

16 The Coriolis effect deflects the flow of air in the northern hemisphere to the right, while in the southern hemisphere it deflects it to the left.

17 Heat energy is circulated from the equator to the poles according to the tri-cellular model. The sun directly heats the air at the equator causing the warm air to rise and diverge towards the poles (Hadley cell). This air starts to sink again between 30° N and S of the equator and moves away from the high pressure towards the 60° N and S point (Ferrel cell) where the air starts to rise again at the polar front (polar cell).

18 Air is forced to rise over a physical barrier such as mountains. As the air is forced to rise, it cools and condenses, clouds form and precipitation happens at the top of the hill.

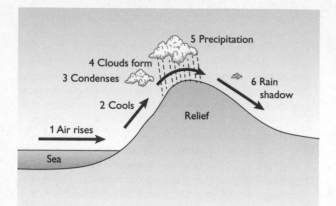

Orographic/relief rainfall

Air above the surface of the Earth is warmed by the direct heating of the Sun. This causes the air to rise, it cools and condenses and clouds form, leading to precipitation.

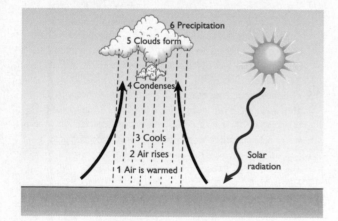

Convectional rainfall

This happens when two bodies of air (air masses) meet. They meet at a front and the air becomes unstable as warm and cold air cannot mix together, so the warm air starts to rise up over the colder dense air. As this happens, the air cools and condenses, clouds form and precipitation takes place.

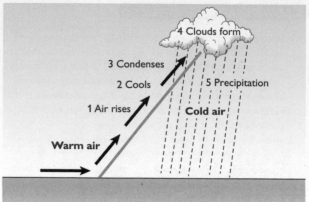

Frontal/cyclonic rainfall

19 The main air mass that affects the British Isles is the tropical maritime air mass. This brings warm and wet conditions and causes dull skies (nimbostratus clouds), drizzle and fog.

20 A depression is an area of low atmospheric pressure that can produce cloudy, rainy and windy weather conditions. Depressions are formed when cold polar maritime and warm tropical maritime air meet in the Atlantic Ocean, creating a weather front. The lighter, warm air starts to rise up over the colder and denser air from the north.

21 As a depression starts to pass overhead the first thing people notice is that the pressure starts to fall and temperatures start to decrease. There is an increase in cloud cover and in wind speeds. As the warm front passes overhead the pressure is low and there is more cloud, with some periods of precipitation. Temperatures increase in the warm sector that follows and skies clear to some extent as calmer conditions develop and less rain falls. However, very quickly temperatures fall as the cold front moves into position, bringing much deeper clouds, as wind speeds increase dramatically and the amount and intensity of rainfall increase. After the cold front has passed conditions remain cold but clouds gradually break up and wind speeds and chances of further precipitation decrease.

22 The main differences between summer and winter anticyclones are mostly to do with temperature and cloud cover and the influences that these bring. In the summer the lack of cloud cover allows temperatures to build, whereas in the winter the lack of cloud cover means that days can be cold, while at night the temperatures can go below freezing, causing frost and icy conditions. The main similarities are that anticyclones usually bring periods of dry, calm, stable weather with few clouds and little precipitation.

23 Hurricanes usually need to form near the equator, in areas where the seawater temperature is above 27°. The airflow needs to start with convectional currents, which allow air to rise and cool quickly in the high levels of the atmosphere. The relative humidity of the air needs to be high so that a massive amount of moisture can be dealt with quickly.